Equity, Diversity, and Inclusion in Healthcare

From Knowledge to Practice

Equity, Diversity, and Inclusion in Healthcare

From Knowledge to Practice

Edited by

Faisal Khosa
University of British Columbia, Vancouver, BC, Canada

Jeffrey Ding
University of British Columbia, Vancouver, BC, Canada

Sabeen Tiwana
University of British Columbia, Vancouver, BC, Canada

ELSEVIER

ACADEMIC PRESS
An imprint of Elsevier

Academic Press is an imprint of Elsevier
125 London Wall, London EC2Y 5AS, United Kingdom
525 B Street, Suite 1650, San Diego, CA 92101, United States
50 Hampshire Street, 5th Floor, Cambridge, MA 02139, United States

ISBN 978-0-443-13251-3

For information on all Academic Press publications
visit our website at https://www.elsevier.com/books-and-journals

Publisher: Mara Conner
Acquisitions Editor: Sonnini Yura
Editorial Project Manager: Sam Young
Production Project Manager: Sujithkumar Chandran
Cover Designer: Christian Bilbow

Typeset by STRAIVE, India

Working together
to grow libraries in
developing countries

www.elsevier.com • www.bookaid.org

Dedication

To my parents for instilling the belief, "anything is possible" and the future generation which holds the potential for endless possibilities: Mahreen, Hamzza, Mariyam, Abdullah, Zamzam, Zunaira, Ahmed, Alaiha, Shahnur, Ahmud, Madinah, Ali, Rania, Mikyle, Nyle, and Ayzel. I would also like to thank the chapter authors, my coeditors, and the Elsevier team for their exemplary support throughout this process. A special thanks to Sonnini Yura, my ever-patient Acquisitions editor.

Endorsements

1. Sanjiv Chopra, MBBS, MACP, FRCP
   Professor of Medicine
   Harvard Medical School
   Editor-in-Chief Hepatology UpToDate
   "The subject of Equity, Diversity, and Inclusion is of paramount importance for all privileged to be in healthcare. In this timely book, every contributor offers a unique perspective. I salute Dr. Faisal Khosa for assembling all of them. I heartily recommend this book to all involved in the unique profession of serving humanity and do so with compassion, grace, humility, and wisdom."

2. Ramneek Dosanjh, MD, CCFP
   President-Elect, Federation of Medical Women of Canada
   President, Doctors of BC (2022–23)
   "Dr. Faisal Khosa has made an exemplary contribution to advancing the need for healthcare equity, diversity, and inclusion. This compelling read should be compulsory for all leaders and participants within the healthcare system. The profound lessons rooted in learning from the healthcare status quo and its deficits can confront injustice in a deliberate and purposeful manner to move toward a more pluralistic society. Disruption can be challenging; however, the walk to a better world is to remind ourselves that healthcare and humanity go hand in hand. Visionary leadership as noted throughout the book is to take the steps required to make an antiracist and inclusive healthcare system that meets the needs of all people, the providers, and the patients. A necessary read and asset to healthcare education!"

3. Alexander Norbash, MD, MA, FACR
   Dean and Professor
   University of Missouri-Kansas City School of Medicine
   "*Equity, Diversity, and Inclusion in Healthcare: From Knowledge to Practice* is an invaluable and transformative resource for healthcare professionals dedicated to fostering inclusive and equitable care across all disciplines. With its comprehensive scope, the book includes essential chapters tailored for specialists in speech and language therapy, optometry, physiotherapy, pharmacy, occupational health, nursing, medicine, dentistry, dietetics, and chiropractic care. Each chapter is meticulously crafted, providing both a deep understanding of the unique challenges and

actionable strategies for implementing effective EDI initiatives in diverse healthcare settings.

What sets this book apart is its practical approach. The authors have skillfully combined theoretical insights with real-world applications, making it an indispensable guide for practitioners at any stage of their EDI journey. Through case studies, best practice examples, and evidence-based recommendations, readers are equipped with the tools needed to create inclusive environments that respect and celebrate diversity, ultimately leading to improved patient outcomes and workplace satisfaction.

As the healthcare landscape continues to evolve, the importance of equity, diversity, and inclusion cannot be overstated. This book is not just a call to action but a roadmap for meaningful change. It challenges professionals to critically assess their practices, embrace cultural competence, and advocate for systemic changes that promote health equity worldwide. For anyone committed to making a tangible difference in the realm of healthcare, this textbook is a must-read, an irreplaceable reference tool, and a 'how to' book that helps every one of us connect the dots between theory and implementation."

4. Kathleen Ross, MD
   President, Canadian Medical Association
   "This exceptional and timely book is an urgent call to action for healthcare professionals, leaders, and advocates to lean into the vital conversation on equity, diversity, and inclusion within the health workforce.

   As each author explores the evidence, the unique, yet overlapping, positive impacts across professional disciplines, comes into sharp focus. We are encouraged to rethink and reshape the future of health by embracing our human diversity. This is intentional, at times difficult work. However, holding space and building trust with others who may not share our same history, interests, or cultural practices encourages personal and professional growth. This process works to realign the power imbalances that limit our best work.

   Our culture of medicine will be stronger when we embrace the full spectrum of diversity in our society, reflect the populations we serve, and foster a sense of belonging within our workforce.
   Enhancing research, training, and practice in settings that uphold the principles of cultural safety, cultural humility, and cultural competency improves outcomes and strengthens the health and wellness of those delivering care, those receiving care, and indeed our ailing healthcare system. The time is now."

5. The Honourable Flordeliz (Gigi) Osler, MD, FRCSC
   "This book acts as an important guide to equity, diversity, and inclusion practices within the healthcare landscape. Authored by a team of leading experts in their profession, the book offers possible solutions to breaking down barriers, stresses the inherent value of diverse perspectives, and recognizes marginalized communities that are underserved by the healthcare system.

Equity, Diversity, and Inclusion in Healthcare: From Knowledge to Practice should be read by healthcare providers, healthcare students, and anyone committed to championing EDI in healthcare."

6. Susan Thompson Hingle MD, MACP, FRCP, FAMWA
   Associate Dean for Human and Organizational Potential
   Southern Illinois University School of Medicine
   President, American Medical Women's Association (2024–25)

   "I applaud the editors for their broad, inclusive lens that breaks down silos in the hierarchical structures of medicine and healthcare and brings together EDI experts with lived experience throughout the healthcare professions, including nursing, pharmacy, medicine, dentistry, physical therapy, chiropractic medicine, nutrition and dietetics, optometry, and speech and language pathology. We are truly better together. In a system where anyone is oppressed, we are all oppressed and unable to reach our full potential as individuals, organizations, or professions. As aboriginal activist, Lilla Watson said, 'If you have come here to help me, you are wasting your time. But if you have come because your liberation is bound up with mine, then let us work together'. The editors offer practical solutions and a roadmap to needed changes that recognize the important impact of diversity, inclusion, and equity on the well-being and the ultimate functioning of our struggling healthcare systems. This book is for anyone interested in building a healthcare workforce that meets the needs of the evolving and diverse global community throughout the world."

7. Richard Brown, DC, LL.M, JP, FRCC, FICC
   Secretary-General
   World Federation of Chiropractic

   "The chapter in this critically important text relating to matters of equity, diversity, and inclusion as they relate to chiropractic is both enlightening and thought-provoking. It provided a comprehensive and well-structured overview of the topic, offering valuable insights and practical guidance on creating more inclusive and diverse environments, for chiropractors, the patients they serve, and wider society.

   The authors have done an exceptional job of explaining the fundamental concepts of equity, diversity, and inclusion, ensuring that readers from various backgrounds can easily grasp their significance. They present a compelling case for why these principles are crucial in today's diverse and interconnected world, emphasizing the moral, social, and economic advantages of embracing diversity and fostering inclusivity.

   One area that stands out is the pragmatic advice that can permit individual chiropractors, chiropractic organizations, and the wider profession in general. This advice helps to contextualize the concepts and demonstrate how equity, diversity, and inclusion can be implemented in different settings.

   Importantly, the chapter did not shy away from addressing the challenges and potential barriers when it comes to the adoption and implementation of principles relating to equity, diversity, and inclusion. The incorporation of actionable strategies and best practices for overcoming these challenges, emphasizing the importance of leadership commitment, cultural competence, ongoing evaluation, and continuous improvement stressed the fact that these matters are an ongoing process of learning, communication, and better understanding, rather than a one-off event. The authors presented their ideas in a clear and concise manner, without oversimplifying or compromising in depth.

In terms of content, the chapter covered a wide range of subjects, reflecting the complexity and multifaceted nature of equity, diversity, and inclusion. It was refreshing to see issues such as hiring and recruitment practices, creating inclusivity in workplace cultures and the building of effective programs being addressed. Each topic was covered in sufficient detail, striking an important balance between theory and practical application.

As a reader, I appreciated the chapter's focus on intersectionality—the recognition that individuals possess multiple social identities that intersect and shape their experiences. This lens added depth and nuance to the discussion, acknowledging the unique challenges faced by individuals with multiple, marginalized identities and emphasizing the need for comprehensive and inclusive approaches.

Overall, this chapter left a lasting impression on me. It provides a valuable resource for both individuals and organizations seeking to foster inclusive environments and embrace the benefits of diversity. I would highly recommend this chapter to anyone interested in understanding and promoting equity, diversity, and inclusion in their personal or professional lives."

Contents

1. Equity, diversity, and inclusion in chiropractic: Aligning the profession to serve tomorrow's diverse world

Claire D. Johnson, Bart N. Green, Sumaya F. Ahmed, Lyndon G. Amorin-Woods, Kara D. Burnham, Waleska Crespo-Rivera, William K. Foshee, Kelley M. Humphries-Mascoll, Craig S. Little, L. David Peeace, Jean-Nicolas Poirier, and Christopher Yelverton

Contributors

Numbers in parenthesis indicate the pages on which the authors' contributions begin.

Mariam M. Abdelaziz (191), Department of Communication Sciences and Disorders, North Carolina Central University, Durham, NC, United States

Sumaya F. Ahmed (1), London South Bank University, London, United Kingdom

Lyndon G. Amorin-Woods (1), Murdoch University, Perth, WA, Australia

Jeffrey John Andrion (173), Department of Physical Therapy, University of Toronto, Toronto, ON, Canada; Department of Physical Therapy; A.T. Still Research Institute, A.T. Still University, Mesa, AZ, United States; Philippine Working Group—International Centre for Disability and Rehabilitation, University of Toronto, Toronto, ON, Canada

Sunny Bhakta (157), Department of Pharmacy Services, Houston Methodist Hospital, Houston, TX, United States

Kara D. Burnham (1), University of Western States, Portland, OR, United States

Gary Y. Chu (139), New England College of Optometry, Boston, MA, United States

Waleska Crespo-Rivera (1), Universidad Central del Caribe, Bayamon, Puerto Rico

Michele A. DeBiasse (47), Department of Health Sciences, Sargent College of Health & Rehabilitation Sciences, Boston University, Boston, MA, United States

Aniela M. dela Cruz (95), Faculty of Nursing, University of Calgary, Calgary, AB, Canada

Sana Dhawan (47), Department of Applied Human Nutrition, Mount Saint Vincent University, Halifax, NS, Canada

Jeffrey Ding (69, 215), University of British Columbia, Vancouver, BC, Canada

Philip R. Doiron (69), Division of Dermatology, Department of Medicine, University of Toronto, Toronto, ON, Canada

Mariyam Durrani (31), Faculty of Medicine, Masaryk University, Brno-Bohunice, Czechia

William K. Foshee (1), MyoCore, Dallas, TX, United States

Bart N. Green (1), National University of Health Sciences, Lombard, IL; Scripps Health, San Diego, CA, United States

Kashif Hafeez (31), Dent Hub Smiles, Gloucester, United Kingdom

Kelley M. Humphries-Mascoll (1), Logan University, Chesterfield, MO, United States

Simone Jadczak (139), New England College of Optometry, Boston, MA, United States

Claire D. Johnson (1), National University of Health Sciences, Lombard, IL, United States

Marissa Joseph (69), Department of Pediatrics, Faculty of Medicine, University of Toronto; Section of Dermatology, Division of Paediatric Medicine, The Hospital for Sick Children; Division of Dermatology, Women's College Hospital, Toronto, ON, Canada

Phillip R. Joy (47), Department of Applied Human Nutrition, Mount Saint Vincent University, Halifax, NS, Canada

Lillian Kalaczinski (139), Michigan College of Optometry at Ferris State University, Big Rapids, MI, United States

Danai Kasambira Fannin (191), Department of Communication Sciences and Disorders, North Carolina Central University; Department of Head and Neck Surgery and Communication Sciences, Duke University School of Medicine, Durham, NC, United States

Faisal Khosa (69, 215), University of British Columbia, Vancouver, BC, Canada

Mahreen Khosa (157), Department of Pharmacy, Mater Private Hospital, Dublin, Ireland

Zamzam Khosa (31), Department of Dentistry, Lahore Medical & Dental College, Lahore, Punjab, Pakistan

Crystal Lewandowski (139), New England College of Optometry; North End Waterfront Health, Boston, MA, United States

Craig S. Little (1), Council on Chiropractic Education, Scottsdale, AZ, United States

Stephanie N. Lurch (173), Department of Physical Therapy, University of Toronto, Toronto; School of Rehabilitation Science, Physiotherapy Program, McMaster University, Hamilton; York Catholic District School Board, Aurora; Children's Treatment Network, Richmond Hill, ON, Canada

Nidhi Mahendra (191), Department of Head and Neck Surgery and Communication Sciences, Duke University School of Medicine, Durham, NC, United States

Ana P.S. Malfitano (119), Occupational Therapy Postgraduate Program, Occupational Therapy Department, Federal University of São Carlos, São Carlos, Brazil

Jairus-Joaquin Matthews (191), Department of Counseling, Higher Education, and Speech Language Pathology, University of West Georgia, Carrollton, GA, United States

Muhammad Mustafa Memon (69), Department of Medicine, Rochester General Hospital, Rochester, NY, United States

Carlos E. Moreno (119), Department of Occupational Therapy, School of Health Professions, Saint Joseph's University, Philadelphia, PA, United States

Efekona Nuwere (119), Occupational Therapy Program, School of Health Professions, SUNY Downstate Health Sciences University, Brooklyn, NY, United States

Jessica A. O'Flaherty (47), Department of Applied Human Nutrition, Mount Saint Vincent University, Halifax, NS, Canada

Richard Ogden Jr. (157), Department of Pharmacy, Children's Mercy Kansas City, Kansas City, MO, United States

L. David Peeace (1), Canadian Indigenous Chiropractic Caucus, SK, Canada

Jean-Nicolas Poirier (1), Northeast College of Health Sciences, Seneca Falls, NY, United States

Chelsey T. Purdy (47), Department of Applied Human Nutrition, Mount Saint Vincent University, Halifax, NS, Canada

Addy Rose (139), New England College of Optometry, Boston, MA, United States

Anthony Scott (157), Emory University Hospital, Atlanta, GA, United States

Zahra Shajani (95), Faculty of Nursing, University of Calgary, Calgary, AB, Canada

Steven D. Taff (119), Program in Occupational Therapy, Department of Medicine, Washington University School of Medicine in St. Louis, St. Louis, MO, United States

Muhammad Haaris Tiwana (31), Faculty of Health Sciences, Simon Fraser University, Burnaby, BC, Canada

Sabeen Tiwana (31, 215), University of British Columbia, Vancouver, BC, Canada

Stacy West-Bruce (119), Program in Occupational Therapy, Washington University School of Medicine in St. Louis, St. Louis, MO, United States

Christopher Yelverton (1), University of Johannesburg, Johannesburg, South Africa

Editors' biographies

Faisal Khosa, MD, MBA, TI, FFRRCSI, FRCPC, FACR
Dr. Khosa is an author, educator, and mentor who strives to attain inclusive and equitable societies that address the overlapping dimensions of discrimination and oppression. His research and advocacy have resulted in tangible improvements and increased access for minorities in educational institutions and employment. The more important yet intangible benefits of this work include the provision of culturally competent care through increasing diversity in the healthcare workforce.

Jeffrey Ding, MD
Dr. Ding is a resident physician in the Department of Family Medicine, University of British Columbia. Jeffrey is an avid EDI researcher with a specific focus on gender, racial, and ethnic disparities in medicine. Jeffrey has published over 30 peer-reviewed articles, with a focus on increasing awareness of workforce disparities and advocating for institutional action within medical schools, medical specialties, professional societies, journal editorial boards, clinical trials, and Canadian health authorities.

Sabeen Tiwana, BDS, DDS
Dr. Tiwana is the owner of a family dental practice in Vancouver and a clinical instructor at the Faculty of Dentistry, University of British Columbia. Sabeen has a keen interest in advocacy and mentoring of students from underrepresented groups, including minorities and refugees. Sabeen is a workplace advisor for several schools, has published extensively on demographic disparities in healthcare professions, and offers courses on bias and harassment.

Land acknowledgment

The editors respectfully acknowledge that the land on which they work and learn is the traditional, ancestral, and unceded territory of the Coast Salish peoples, including the territories of the Musqueam, Squamish, and Tsleil-Waututh Nations. We are grateful for the opportunity to be on this land and to contribute to the ongoing process of reconciliation by promoting equity, diversity, and inclusion in healthcare.

Traditionally, it is expected that one asks for permission before stepping onto someone else's territory. In this spirit, the editors request all readers to reflect on the importance of engaging thoughtfully and respectfully with the history and ownership of the land they reside on.

Introduction: Fostering an age of equity, diversity, and inclusion—A primer from the editors

Faisal Khosa, Jeffrey Ding, and Sabeen Tiwana
University of British Columbia, Vancouver, BC, Canada

The delivery of effective, comprehensive, and culturally competent patient care coincides with ensuring equity, diversity, and inclusion (EDI) in the healthcare professions. As the general patient population trends towards greater diversity, a sustainable and efficient healthcare system can only exist when the providers of care reflect those whom they serve. Acknowledging that the healthcare workforce is an integral pillar of societal wellbeing, it is important to recognize that safeguarding public health must include promoting equitable healthcare. The value of EDI goes beyond a matter of political correctness. Rather, nurturing equity within the healthcare workforce is the first step in fostering a compassionate, robust, and patient-centric ecosystem. The workforce effectively stands as the bedrock for this complex healthcare network. Only once EDI is embraced at the systemic level will we witness the downstream benefits of improved health outcomes and patient experiences.

While the promotion of EDI as a collective term continues to become a topic of increasing popularity in recent years, the individual words of "equity," "diversity," and "inclusion" should be understood and defined clearly. The definitions we use in our work are as follows, adapted from (Corbie et al., 2022). "Equity" refers to the fair provision and distribution of resources and power, so that the full potential may be realized by all individuals. To achieve equity, we must eliminate the privilege and unearned advantage of historically included groups, and the disparities, unearned disadvantages, and oppression of historically excluded groups. "Diversity" refers to the representation and mix of identities, similarities, and differences (both on the individual and collective levels). Examples of such demographics include gender identity, ethnicity, race, disability, sexual orientation, personal historical experiences, and socioeconomic status, among others. "Inclusion" refers to the conceptual state by which all people, perspectives, and voices can contribute and be heard. For their talents to be utilized, all individuals and groups must be allowed for full participation, a sense of belonging—true engagement is not merely a seat at the table.

Historically, those from underrepresented and marginalized populations have faced barriers in gaining representation within the healthcare professions. These disparities have further been accentuated among those holding leadership positions or roles which granted power and influence. It is challenging to kick-start a movement to foster greater EDI without support from the upper echelons of the workforce, and therefore, the reality is that we have waited until the 21st century for the EDI movement to swing into motion and to gain momentum. Factors such as gender, race, ethnicity, socioeconomic status, physical disability, and sexual orientation are only a few out of the endless categories which EDI aims to address. EDI is constantly evolving, expanding, and encompassing new groups. For instance, the acronym LGBTQIA2S+ acknowledges several gender and sexual identities (L: lesbian, G: gay, B: bisexual, T: transgender, Q: queer/questioning, I: intersex, A: asexual, 2S: two-spirit), while reserving space for new groups and individuals who do not fit in the existing mold (+: inclusive of those who exist beyond the established categories within this community).

The impetus for improving EDI in the healthcare workforce can be rationalized with evidence from the literature. In modern society where populations continue to densify in urban hubs, there is a clear tendency for healthcare resources and personnel to congregate in these locations. However, this inevitably creates underserved communities, most notably those in rural areas. Research has demonstrated that women and those from ethnic/racial minority backgrounds are more likely to work in underserved communities (Garcia et al., 2018; Komaromy et al., 1996; Moy & Bartman, 1995; Rabinowitz et al., 2000; Xierali et al., 2014). Research has also shown that a physician's willingness to provide care in these remote/underserved communities is best predicted by race and ethnicity (Lightfoote et al., 2014; Xierali & Nivet, 2018). In addition to improving access to healthcare, a diverse workforce also bridges cultural differences, provides culturally sensitive care, and reduces language barriers. Patient-centered care starts with fostering trust through creating a provision of culturally competent care providers. Patient-provider relationships can be enhanced when social and/or cultural backgrounds match, which facilitates greater understanding and disclosure of health concerns (Ding et al., 2023; Yong-Hing & Khosa, 2023).

As Desmond Tutu put it, "True reconciliation is never cheap, for it is based on forgiveness which is costly. Forgiveness in turn depends on repentance, which has to be based on an acknowledgment of what was done wrong, and therefore on disclosure of the truth. You cannot forgive what you do not know." The impetus for raising awareness of EDI issues in the healthcare workforce comes from a need to generate meaningful discourse between various stakeholders. However, this dialogue must be built on a transparent understanding of the history of systemic inequities, which is challenging to accomplish without literature that is comprehensive and all-encompassing. Our work aims to address this tangible gap in the literature. Acknowledgement of historical injustices is of paramount importance to move towards EDI. Without this fundamental step of documentation and information dissemination, there cannot be an understanding of the full scope of existing and historical disparities within the healthcare workforce. Only when transparency and accountability are ensured can there be healing, reconciliation, and development of targeted policies and interventions that address the specific needs of marginalized groups.

A focused discussion on disparities at the leadership level is important to drive change among the greater workforce population. We must move beyond the "think leadership, think white male" mindset (Soklaridis et al., 2022). We must move away from the paradigm of thinking that "best person for the job" has anything to do with the circumstances in which an individual is born. The reality is that the best people for the job are made—rather than born—through access to strategically important information, coaching, sponsorship, and connections to influential networks (Frank, 2022). The disparity lies in that those from underrepresented backgrounds do not have the same access and resources. Having representation at the highest levels of leadership can foster a sense of belonging and promote measures such as mentorship initiatives, inclusive hiring practices, and leadership development programs. These initiatives can then benefit the next generation of underrepresented individuals by equipping them with the necessary toolkits to make them into the "best person for the job."

It has been proposed that a new direction for leadership development to advance health equity should be through an "equity-centered" framework. Instead of regarding EDI principles as a handful of add-on or checkbox competencies, these should be assimilated fully into the core/traditional leadership requirements. Corbie and colleagues outlined four fundamental domains in their equity-centered leadership framework which synergizes contemporary EDI skills and traditional leadership skills (Corbie et al., 2022). The "personal" domain encourages equity through the cultivation of self-awareness and appreciation of individual differences to facilitate engagement with diversity. The "interpersonal" domain focuses on inclusion by developing communication skills/conflict management and calling in through the practice of multiculturalism. The "organizational" domain aims to shift organizational and leadership culture to facilitate a work environment where all members feel engaged, valued, and functioning at their full potential. At last, the "community and systems" domain pertains to developing skills required to address drivers of health equity in a collaborative approach with communities and stakeholders. In essence, leadership development is the crux of promoting health equity for generations to come. As the times change, the necessary skillsets for leaders also shift. Traditional leadership competencies must be expanded to fully integrate EDI principles to safeguard equitable opportunities for all.

The discussion surrounding inequity should not be exclusive to the underrepresented and less privileged. Whenever there is a social structure which creates disadvantages for certain groups, we must also pay attention to those with unearned and unfair advantages. Nixon introduced the coin model of privilege and critical allyship, which outlines an intersectional approach to understanding how the components of inequality such as sexism, racism, and ableism generate oppression and privilege (Nixon, 2019). In this model, "coins" are used to represent these components of inequality (e.g., sexism, racism, and ableism), which are seen to have two sides. The top of the coin denotes the groups who benefit from privilege and unearned advantages. The bottom of the coin represents the marginalized and disadvantaged groups. The key message is not to find interventions which will transition people from "below the coin" to "above the coin," because both the sides are inequitable and contributing to the problem of health disparities. Health inequity is often focused only on the bottom of the coin; however, how you frame the problem also determines the scope of which the solutions target. Positions of unearned privilege and advantage must come under equal scrutiny to dismantle the systemic drivers of inequities in the healthcare professions.

To address inequities and disparities, we must be pragmatic and not analyze one system of oppression at a time. As Dr. Martin Luther King said, "Injustice anywhere is a threat to justice everywhere. We are caught in an inescapable network of mutuality, tied in a single garment of destiny. Whatever affects one directly, affects all indirectly." The interplay of these various dimensions of our identity are inadequately captured when the intricacies of human experiences are cross-sectionally snapshotted through individual facets of oppression. Marginalization and discrimination cannot be understood fully without a holistic approach through multisystem analyses. For example, gender, race, and sexual orientation are experienced simultaneously but seldom analyzed concurrently. With respect to the coin model, individuals can be on the top of some coins while simultaneously occupying the bottom of various other coins (Nixon, 2019). The complexity lies in the fact that privilege is not simply additive. An intersectional approach acknowledges that there are several layers of discrimination experienced simultaneously, and therefore, we cannot promote EDI and amplify the voices of the marginalized without recognizing the complex interactions between unique dimensions of identity.

The principle of EDI is not to work in isolation between the marginalized and nonmarginalized communities. The purpose of this work is to create a workforce that is safe for all, hence the pillar of "inclusion" in EDI. Furthermore, the analysis of advantages and disadvantages as it pertains to healthcare equity is not about attributing innocence or guilt (Nixon, 2019). Those who find themselves in positions of privilege or unearned advantage should not feel like their role in the EDI movement is less pivotal than those facing unfair disadvantages. This is where critical allyship comes into play. An individual does not simply "identify" as an ally, but rather they must engage in practices that immerse themselves as a partner in a collaborative effort to address inequities. The Anti-Oppression Network defines allyship as "active, consistent, and arduous practice of unlearning and re-evaluating in which a person of privilege seeks to operate in solidarity with a marginalized group of people" (Allyship, 2015). The flawed way of thinking as a privileged individual is that the disadvantages faced by marginalized communities are simply "their problems." The word "ally" therefore is a verb. Instead, these are everyone's problems, but often those who reap the benefits choose not to recognize their role in these systems of inequality. As suggested by Nixon, engagement in critical allyship requires a reassessment of the way we think. For example, instead of subscribing to the mindset of "I wish to help the less fortunate" or "I use my expertise to reduce inequities for marginalized populations," one should reorientate their thinking to "I seek to understand my own role in upholding systems of oppression that create health inequities" and "I learn from the expertise of, and work in solidarity with, historically marginalized groups to help me understand and take action on systems of inequality" (Nixon, 2019).

Building a healthcare workforce that meets the needs of the evolving and diverse global community requires recognizing the significance of EDI. This is not an elusive concept that benefits only a select few, but rather it is a fundamental movement which has implications for emotional, psychological, and sociocultural wellbeing. For the provision of sustainable, efficient, and compassionate healthcare, there must be a concerted effort on the individual, organization/institutional, and societal levels. Furthermore, to make certain that the whole is greater than the sum of its parts, there needs to be a cohesive effort to improve EDI across the healthcare professions. As such, our work is a collaboration between EDI experts with lived experience in healthcare professions including chiropractic medicine, dentistry, dietetics, medicine, nursing, optometry, pharmacy, physiotherapy, and speech language pathology. Each chapter characterizes the present state of workforce diversity with a review of the longitudinal trends in EDI developments, and finally present the rationale for systemic action, accompanied by solutions, interventions, and possible initiatives to tackle such disparities. By bringing together EDI experts from the healthcare professions, we hope to usher in an age of greater EDI.

The most common explanation by leaders and those in positions of influence is that equity, diversity, and inclusion take time. My question is, whose time – that of a human or a God? The scriptures tell us that God's one day is equal to one thousand human years. So, if we are talking in terms of God's time, it has only been six days. If we are talking in terms of human time, then recorded human history is six thousand years. How much longer are we expected to wait?

Faisal Khosa

References

Allyship. (2015). *The anti-oppression network*. https://theantioppressionnetwork.com/allyship/.

Corbie, G., Brandert, K., Fernandez, C. S. P., & Noble, C. C. (2022). Leadership development to advance health equity: An equity-centered leadership framework. *Academic Medicine, 97* (12), 1746–1752. https://doi.org/10.1097/ACM.0000000000004851.

Ding, J., Yong-Hing, C. J., Patlas, M. N., & Khosa, F. (2023). Equity, diversity, and inclusion: Calling, career, or chore? *Canadian Association of Radiologists Journal*, *74*(1), 10–11. https://doi.org/10.1177/08465371221108633.

Frank, T. J. (2022). *The Waymakers: Clearing the path to workplace equity with competence and confidence*. Amplify Publishing. https://www.amazon.ca/Waymakers-Clearing-Workplace-Competence-Confidence/dp/1637551800.

Garcia, A. N., Kuo, T., Arangua, L., & Pérez-Stable, E. J. (2018). Factors associated with medical school graduates' intention to work with underserved populations: Policy implications for advancing workforce diversity. *Academic Medicine*, *93*(1), 82–89. https://doi.org/10.1097/ACM.0000000000001917.

Komaromy, M., Grumbach, K., Drake, M., Vranizan, K., Lurie, N., Keane, D., & Bindman, A. B. (1996). The role of black and Hispanic physicians in providing health care for underserved populations. *New England Journal of Medicine*, *334*(20), 1305–1310. https://doi.org/10.1056/NEJM199605163342006.

Lightfoote, J. B., Fielding, J. R., Deville, C., Gunderman, R. B., Morgan, G. N., Pandharipande, P. V., Duerinckx, A. J., Wynn, R. B., & Macura, K. J. (2014). Improving diversity, inclusion, and representation in radiology and radiation oncology part 2: Challenges and recommendations. *Journal of the American College of Radiology*, *11*(8), 764–770. https://doi.org/10.1016/j.jacr.2014.03.008.

Moy, E., & Bartman, B. A. (1995). Physician race and care of minority and medically indigent patients. *JAMA*, *273*(19), 1515–1520.

Nixon, S. A. (2019). The coin model of privilege and critical allyship: Implications for health. *BMC Public Health*, *19*(1), 1637. https://doi.org/10.1186/s12889-019-7884-9.

Rabinowitz, H. K., Diamond, J. J., Veloski, J. J., & Gayle, J. A. (2000). The impact of multiple predictors on generalist physicians' care of underserved populations. *American Journal of Public Health*, *90*(8), 1225–1228. https://doi.org/10.2105/ajph.90.8.1225.

Soklaridis, S., Lin, E., Black, G., Paton, M., LeBlanc, C., Besa, R., MacLeod, A., Silver, I., Whitehead, C. R., & Kuper, A. (2022). Moving beyond 'think leadership, think white male': The contents and contexts of equity, diversity and inclusion in physician leadership programmes. *BMJ Leader*, *6*(2), 146–157. https://doi.org/10.1136/leader-2021-000542.

Xierali, I., Casillo-Page, L., Conrad, S., & Nivet, M. (2014). Analyzing physician workforce racial and ethnic composition associations: geographic distribution (part II). In *Vol. 14. Analysis in brief* AAMC. https://www.aamc.org/media/7621/download.

Xierali, I. M., & Nivet, M. A. (2018). The racial and ethnic composition and distribution of primary care physicians. *Journal of Health Care for the Poor and Underserved*, *29*(1), 556–570. https://doi.org/10.1353/hpu.2018.0036.

Yong-Hing, C. J., & Khosa, F. (2023). Provision of culturally competent healthcare to address healthcare disparities. *Canadian Association of Radiologists Journal*, *74*(3), 483–484. https://doi.org/10.1177/08465371231154231.

Chapter 1

Equity, diversity, and inclusion in chiropractic: Aligning the profession to serve tomorrow's diverse world

Claire D. Johnson[a], Bart N. Green[a,b], Sumaya F. Ahmed[c],
Lyndon G. Amorin-Woods[d], Kara D. Burnham[e], Waleska Crespo-Rivera[f],
William K. Foshee[g], Kelley M. Humphries-Mascoll[h], Craig S. Little[i],
L. David Peeace[j], Jean-Nicolas Poirier[k], and Christopher Yelverton[l]

[a]National University of Health Sciences, Lombard, IL, United States, [b]Scripps Health, San Diego, CA, United States, [c]London South Bank University, London, United Kingdom, [d]Murdoch University, Perth, WA, Australia, [e]University of Western States, Portland, OR, United States, [f]Universidad Central del Caribe, Bayamon, Puerto Rico, [g]MyoCore, Dallas, TX, United States, [h]Logan University, Chesterfield, MO, United States, [i]Council on Chiropractic Education, Scottsdale, AZ, United States, [j]Canadian Indigenous Chiropractic Caucus, SK, Canada, [k]Northeast College of Health Sciences, Seneca Falls, NY, United States, [l]University of Johannesburg, Johannesburg, South Africa

---

We envision a world where all people attain the highest possible standard of health and well-being; where diversity of all kinds is celebrated; human rights are promoted, protected, and fulfilled; gender equality and health equity are the norm; and barriers to health and well-being are addressed.

World Health Organization (2023)

Chiropractic workforce diversity

The limited diversity in the educational institutions and healthcare workforce is an impediment toward the provision of high-quality, culturally competent care to an increasingly diverse world population. This lack of diversity also leads to lack of role models and mentors for students from underrepresented and minority groups. Improvements in these areas are crucial if our world is to have a healthier future.

There are more than 100,000 practitioners in the chiropractic profession worldwide (Stochkendahl et al., 2019). Chiropractors practice in over 100 countries, and chiropractic services are incorporated in public and national

Equity, Diversity, and Inclusion in Healthcare. https://doi.org/10.1016/B978-0-443-13251-3.00001-6

healthcare systems (Field & Newell, 2016; Weigel et al., 2014). According to the World Federation of Chiropractic (WFC), chiropractic is "A health profession concerned with the diagnosis, treatment and prevention of mechanical disorders of the musculoskeletal system, and the effects of these disorders on the function of the nervous system and general health. There is an emphasis on manual treatments including spinal adjustment and other joint and soft-tissue manipulation" (World Federation of Chiropractic, 2001).

As evidence-based healthcare providers, chiropractors incorporate best evidence, clinical expertise, and patient values in their practices (Himelfarb et al., 2020). The incorporation of patient values is an essential component to providing person-centered care, which occurs when the provider respects an individual's culture, beliefs, and values, thereby empowering people to participate in the healing process (Johnson et al., 2018). Effective person-centered care requires communication, shared decision-making, and holistic approaches, which address the whole person and not only their biomedical complaint (Johnson et al., 2018).

Chiropractic care includes conservative healthcare methods that emphasize the diagnosis, treatment, and prevention of neuromusculoskeletal disorders and their effects on general health (Brown, 2016; Christensen et al., 1993, 1994; Coulter & Shekelle, 2005; General Chiropractic Council, 2004; Glucina et al., 2019; Himelfarb et al., 2020; Humphreys et al., 2010; Imbos et al., 2005; Johl et al., 2017; Pollentier & Langworthy, 2007; Puhl et al., 2015). Chiropractors use a biopsychosocial approach to care, incorporating concepts of body, mind, and social interactions when treating patients. In addition to manual techniques (i.e., joint manipulations, soft tissue mobilization) and rehabilitation, chiropractors provide to patients education and advice for prevention and self-care (Haldeman, Johnson, et al., 2018). Chiropractors recognize the importance of provider-patient rapport and relationships with their patients, resulting in high patient satisfaction (Crowther, 2014; Gaumer, 2006; Gemmell & Hayes, 2001; Rowell & Polipnick, 2008; Sawyer & Kassak, 1993; Weigel et al., 2014). This includes equity, diversity, and inclusion (EDI). The chiropractic profession recognizes the value of EDI. The WFC is a nongovernmental organization in official relations with the World Health Organization and is a global not-for-profit organization that represents national chiropractic organizations. The EDI policy of the WFC includes promotion of equality, fairness, and respect for all in our global community (World Federation of Chiropractic, 2020).

Chiropractic education and EDI

During commencement or when entering clinical placements, chiropractors take an oath or make a pledge (Chiropractic Educators Research Forum, 2022a; Deltoff, 2023; Deltoff & Deltoff, 1988; Simpson et al., 2010). With this oath, they declare that they will serve humanity in the patient's best interest without discrimination and will do their best to help all who seek their care

(Deltoff, 2023). A duty to others requires that chiropractors are effective in their ability to serve and, therefore, understand patients' needs and provide appropriate care. As such, empathy, cultural humility, and commitment to serving the public are professional competencies that are expected of chiropractors. In other words, equity and inclusion are requisite to what chiropractors do.

A great deal of recent work in EDI in chiropractic has transpired in education. Throughout the world, chiropractors are trained with EDI competencies in mind. Globally, chiropractic education accrediting agencies require that graduates demonstrate cultural competency. The accrediting agencies for various world regions include the Council on Chiropractic Education (United States [US] and Canada), the Council on Chiropractic Education Australasia (Australia, New Zealand, Malaysia), the European Council on Chiropractic Education (United Kingdom [UK], Spain, South Africa, France, Denmark, Wales, Switzerland), and the Federation of Canadian Chiropractic (Canada).

The Council on Chiropractic Education 2025 standards include that students must be able to "demonstrate an awareness of biases and social determinants of health that may impact the delivery of care to a diverse population; evaluate the role of sociocultural, socioeconomic, and diversity factors in contemporary society to meet the healthcare needs of persons, groups, and populations; graduates must be able to communicate respectfully and effectively with patients of diverse social, cultural, and linguistic backgrounds in a manner that protects the dignity of individuals and communities; and, graduates must be able to design a care plan that considers and respects the culture of the patient" (Council on Chiropractic Education, 2024). Also included in the US chiropractic accreditation standards is a metacompetency that states "Cultural competency includes the knowledge, skills, and core professional attributes needed to provide care to patients with diverse values, beliefs, and behaviors, including the tailoring of health care delivery to meet patients' social, cultural, and linguistic needs in an effort to reduce disparities in healthcare delivery" (Council on Chiropractic Education, 2024).

The Council on Chiropractic Education Australasia 2017 standards include that graduates must be able to recognize and respond "to diversity in the population, including but not limited to gender, age, religion, race, disability, socioeconomic status, and sexual orientation; and recognize and respond to the impact of culture, values, beliefs, education levels and life experiences on health status, health and help-seeking behaviors and maintenance of health" (Council on Chiropractic Education Australasia, 2017).

The Federation of Canadian Chiropractic 2018 standards state that chiropractic programs must prepare their graduates to "engage in responsive, nonjudgmental, and culturally respectful dialogue, during written (including electronic) communication, verbal, and non-verbal communication; demonstrate respect for patient, family, and community cultural and social values in the provision of clinical care; and adapt to a variety of patient types and populations" (Federation of Canadian Chiropractic, 2018).

The European Council on Chiropractic Education standards underscore all of the above, including that graduate chiropractors will have knowledge and understanding of the varying cultural, gender, and ethnic differences of patients (European Council on Chiropractic Education, 2019). Thus, the world regions where chiropractic has the greatest presence require EDI competencies for training.

Although the profession's standards reflect the value of EDI, chiropractic educators have expressed the need for greater awareness and dedicated action to support EDI within the profession (Callender, 2006; Chiropractic Educators Research Forum, 2022b; Johnson et al., 2021; Johnson & Green, 2012; Johnson, Killinger, et al., 2012; Southerst et al., 2022; Young, 2015). Despite this concern, few studies have described outcomes or initiatives to address the issue of disproportional representation within the profession globally. Useful data are available from various reports, such as practice analyses. These studies are beneficial to monitoring demographic traits of chiropractors over the years; however, these are published infrequently, are heterogeneous among world regions, and do not contain data on all countries where chiropractors are licensed (Christensen et al., 1993, 1994; Himelfarb et al., 2020; Humphreys et al., 2010; Imbos et al., 2005; Johl et al., 2017). The four countries with the longest existing chiropractic programs and greatest number of chiropractors are the US (75%), Canada (8%), Australia (5%), and the UK (3%), comprising more than 90% of the global workforce of chiropractors (Johnson et al., 2022; Stochkendahl et al., 2019). Because these countries have the resources to prepare demographic reports, the findings tend to be focused on these regions. More research and demographic reports are needed to better understand global representation.

Sex, gender, and sexual orientation diversity

Sex is typically assigned at birth and refers to the biological makeup of a human in terms of chromosomes, hormones, and primary and secondary sex characteristics (Pinn, 2003; Torgrimson & Minson, 2005).

The self-reported sex distribution of chiropractors by country is shown in Fig. 1. The US has the most unequal representation of sex diversity of chiropractors with a 1:2.1 ratio of women to men (Himelfarb et al., 2020). This is likely because chiropractic has existed in the US for the longest time. More chiropractors in this cohort are over 50 years of age, which means they likely graduated at a time when women were less likely to choose chiropractic as a career. Australia's women to men ratio is 1:1.4 and Canada's ratio is 1:1.2, whereas the UK has a 1:1 ratio (Chiropractic Board of Australia, 2023; General Chiropractic Council, 2021; Southerst et al., 2022). These findings may be because chiropractic is a newer profession in these regions. The more balanced ratios may also reflect the healthcare culture, including the recruiting methods used for the chiropractic programs in these countries, which seems more gender neutral.

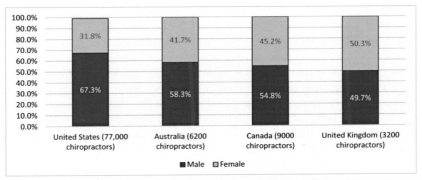

FIG. 1 Representation of women and men in the four countries with the greatest proportion of chiropractors worldwide.

In addition, the chiropractic student population is trending toward a more representative ratio of women to men. In some countries, such as South Africa, the student ratio favors women (Yelverton et al., 2022).

Looking at North American and international chiropractic groups that focus their practices on specific populations, such as veterans' health or sports medicine, the ratio of women to men shows a substantially greater proportion of male practitioners (1:2.9–1:3.5) (Fig. 2) compared to the typical chiropractic population (Fig. 1) (Nelson et al., 2021). It is possible that there are some pre-existing biases or barriers in these historically male-dominated practice environments, which may have discouraged women from entering these special interest healthcare areas.

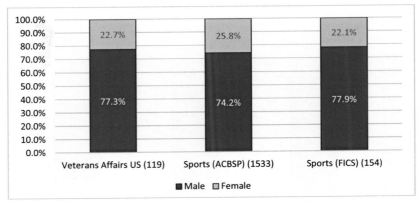

FIG. 2 Sex of chiropractors who work in integrated healthcare systems, such as US Veterans Affairs locations and chiropractors who are listed on the roster with either the American Chiropractic Board of Sports Physicians (ACBSP) or the International Federation of Sports Chiropractic (FICS).

The history of the chiropractic profession has influenced its demographics. In the 1800s when other healthcare professions did not allow women to matriculate, chiropractic programs welcomed women as students and professors. Thus, women have been part of chiropractic since its inception in the U S in the late 1800s. However, although women freely matriculated and graduated from chiropractic programs, most graduates were men, which may have been due to the social norms of that time. Chiropractic practice requires a substantial amount of time to run a practice, which likely competed with the then-traditional role of women being mothers and housekeepers during the 19th and 20th centuries. The physical attributes and technical standards required to perform chiropractic treatments (e.g., spinal manipulation and manual treatments) and the substantial amount of physical contact with patients may have been perceived as physical or cultural barriers for some women. In addition, chiropractic requires psychomotor skills and physical fitness, which in some cultures have been attributed to being a man. However, the myth that men are superior to delivering the force required for chiropractic manipulation has been disproven with a study showing that women are as effective as men in generating forces needed for chiropractic treatments (Forand et al., 2004).

It may be that the trend of increasing numbers of female chiropractic students is becoming reflected in the chiropractic workforce. For example, over the past several decades, there has been a trend toward equal proportions of male and female chiropractors in the US (Fig. 3) (Himelfarb et al., 2020). It is interesting to note that this trend defies that of a decreasing number of women in the US workforce over the past 20 years (Machovec, 2023). A more equal representation of women in the chiropractic workforce may be beneficial for several reasons. In many regions, more patients who seek chiropractic care are women (Brown et al., 2014; Coulter, 1985; Hawk et al., 2000). In addition,

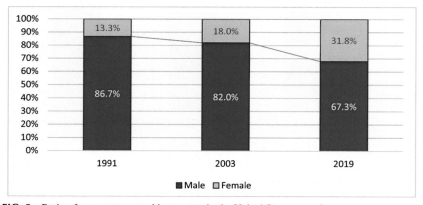

FIG. 3 Ratio of women to men chiropractors in the United States over time.

women healthcare providers are known to demonstrate professional behaviors and actions that are different from men, such as tending to spend more time with patients during patient care (Vollenweider et al., 2017).

Sex and gender are distinctly different terms, where gender refers to "social and cultural influences based on sex" (Pinn, 2003). Sexual orientation is a person's sexual attraction to other people, whereas gender identity is the psychological sense of self, despite the sex a person was assigned at birth (Human Rights Campaign, n.d.). These traits influence communications and are a substantial component of some cultures, which in turn influences the healthcare environment. Thus, gender and sexual identity are important factors when considering the doctor-patient relationship in chiropractic practice (Lady & Burnham, 2019; Maiers et al., 2017).

Even though sex and gender are important identity traits, more population data for these traits within the chiropractic profession are needed. It is possible that due to a lack of awareness, legal issues, or political pressure, information about sexual orientation and gender has not routinely been collected in detail in social and demographic research on the chiropractic profession. For example, the most recent practice analysis performed by the National Board of Chiropractic Examiners expanded demographic questions to include sex and gender (Himelfarb et al., 2020). Given that a portion of the adult population reports to be nonheterosexual in the US (7%), Canada (4%), Australia (3%), and UK (3%), more awareness of these diversity traits is needed (Jones, 2022; Office for National Statistics, 2023; Statistics Canada, 2024; Wilson et al., 2020). If greater awareness and value for this trait are to occur, these data need to be collected, measured, and reported.

Racial and ethnic diversity

The race and ethnic makeup of patients and health providers influences the care provided (Fiscella & Sanders, 2016; Yong-Hing & Khosa, 2023). It has been proposed that healthcare consumers gravitate toward providers with a similar race (Garcia et al., 2003; Godager, 2012; Moore et al., 2023).

Measuring racial and ethnic representation globally in chiropractic is challenging because some regions do not collect this information. Furthermore, characteristics that may be considered a racial or ethnic "minority" in one region may be a majority in another region, even within the same country. How race is defined also varies by region. Since race is a social construct, it is nearly impossible to compare racial group proportions among regions. However, from the information available from the US, Canada, and the UK, it can be interpreted that the chiropractic workforce comprises the races/ethnicities shown in Fig. 4.

At present, chiropractors show a white majority, which is not reflective of the race/ethnicities of the population of these countries (General Chiropractic Council, 2022; Himelfarb et al., 2020; Southerst et al., 2022). There are hypotheses for why the racial makeup is predominately white. One hypothesis is that

United States	Canada	United Kingdom
• White 90.8%	• White 80.3%	• White British 63%
• Black 1.6%	• Black .52%	• White other 1%
• Hispanic/Latino 3%	• Hispanic/Latino .33%	• Mixed 1%
• Asian/Pacific Islander 2%	• Asian/Pacific Islander 12.5%	• Asian or Asian British 3%
• Native American .9%	• Native American .16%	• Black or Black British 1%
• Other or multi 1.7%	• Other or multi 6.3%	• Chinese 1%
		• Not reported 31%

FIG. 4 Reported percent race/ethnicity of the chiropractic workforce for countries with the largest number of chiropractors.

chiropractic initially developed in the US and, therefore, has faced the same struggles with diversity related to race and ethnicity as that of US educational institutions and healthcare professions (Marzbanrad et al., 2024; Xu et al., 2024). In the past century, similar to other healthcare professions (Baker et al., 2008; Zhu et al., 2021), a few chiropractic programs were racially biased, whereas other programs did not impose racial and ethnic barriers (Callender, 2006; Wiese, 2003). The lack of diversity, mentors, role models, and racial biases from the past continue to impact those who enter the profession today.

For many of the profession's early years, chiropractic programs were privately owned and subsisted primarily upon student tuition (Johnson & Green, 2021b). Thus, only matriculating students with the means to acquire adequate funding to complete a private education could be admitted. Throughout the 1900s, nearly all chiropractors were in private practice, which also required a substantial financial outlay. Those with financial privileges were therefore more likely to have the capacity to pay for tuition and afford setting up practices. In addition, patients who were not able to afford care were less likely to be exposed to chiropractic care and, therefore, less likely to consider chiropractic as a career option. Thus, people from socioeconomically disadvantaged groups faced barriers to entering the profession.

Other considerations regarding race and ethnicity are general socioeconomic injustices prevalent in various countries and how they affect healthcare. In North America, Black, Asian, Hispanic/Latinx, Indigenous, and other racially underrepresented communities have been subject to social injustices. Examples include unethical medical experimentation, internment camps, genocide, and barriers to healthcare resources (Armstrong et al., 2007; Bastos et al., 2018; Blakemore, 2018; Brandt, 1978; Freimuth et al., 2001; Gamble, 1993; Moscou et al., 2024; Ojo-Aromokudu et al., 2023; Renzaho et al., 2013; Shamoo, 2023). These events and environments have contributed to inequitable treatment, hindering of human rights, unfair allocation of resources, and poorer health outcomes (Armstrong et al., 2007; Bastos et al., 2018; Moscou et al., 2024; Ojo-Aromokudu et al., 2023; Renzaho et al., 2013). All of these have resulted in people in underrepresented or marginalized groups distrusting healthcare providers within the established healthcare system.

Although chiropractors were not responsible for past racial experiments or other atrocities, they are still considered as members of the established healthcare system. Anyone participating within the healthcare system, particularly those who are from dominant racial and ethnic groups, may be viewed as untrustworthy by people from underrepresented communities (Shippee et al., 2012; Stevens, 2007a). This may result in fewer students from underrepresented cultures finding an interest in pursuing a career in healthcare, including chiropractic. Role models are potent influencers for underrepresented youth when choosing a career (Bonifacino et al., 2021; Callender, 2006; Johnson et al., 2021; Stronach et al., 2023; Wiese, 2003). Thus, racial and ethnic representation within healthcare providers is essential to future recruitment of racially and ethnically diverse student populations.

Chiropractic demographics in the US are changing slowly (Fig. 5) (Himelfarb et al., 2020). Yet, there are few studies about people from racially and ethnically diverse backgrounds who have sought out careers in chiropractic. One small study described the interests and backgrounds of diverse people who chose a chiropractic career (Johnson et al., 2021). Irrespective of their race or ethnicity, they chose the chiropractic profession for a variety of reasons, which included a personal experience in which they benefitted from chiropractic including but not limited to a thirst for knowledge, and a desire to serve others. Qualitative studies such as these may help to better understand how to recruit more racially and ethnically diverse students to chiropractic programs.

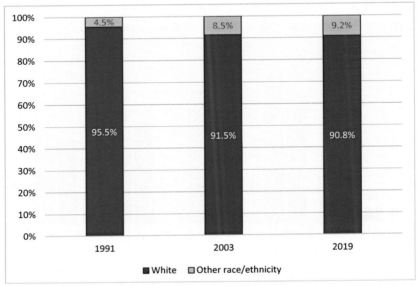

FIG. 5 The gradual change in race/ethnicity of chiropractors in the United States (77,000 chiropractors).

Age diversity

Age as a diversity trait impacts relationships and influences intergenerational communications between providers and patients. The age distribution among chiropractors is somewhat influenced by how long the profession and chiropractic degree-granting programs have been recognized in each country. The U S, having the most graduates over the longest time, has a larger percentage of older practicing chiropractors compared to other countries (Fig. 6A and B). These data are influenced by career longevity; many chiropractors practice for four or more decades.

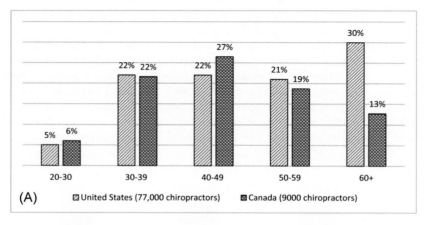

(A) ▨ United States (77,000 chiropractors) ▨ Canada (9000 chiropractors)

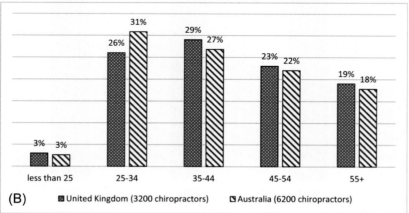

(B) ▨ United Kingdom (3200 chiropractors) ◣ Australia (6200 chiropractors)

FIG. 6 (A and B) Age distribution of chiropractors in the United States, Canada, United Kingdom, and Australia. The countries are reported as two separate groups because the data are collected using different scales.

Other diversity traits

Diversity represents more than race, age, and sex. Diversity includes a combination of individual identity factors, such as religion, socioeconomic status, gender, sexual preference, language, education, abilities (mental, physical), and geographic background. While data about the geographic location of chiropractors can be identified, information regarding other traits is not typically collected (MacMillan et al., 2022; Palmgren et al., 2013).

For example, many chiropractic educational program course catalogs show that they accommodate chiropractic students who are impaired either visually (i.e., partially or fully), mentally (e.g., mental health issues, anxiety), cognitively (e.g., cognitive differences such as dyslexia), or physically (missing limbs). However, little literature is available to explain how these accommodations are made (Joshi & Ray, 2019). The prevalence of these traits in students and practitioners throughout the profession is unknown. It is possible that training program policy may influence what types of traits someone must have to qualify to be a chiropractor. Regional legislation may also affect the kind of data that are allowed to be reported. In some regions, some traits may be considered taboo or not acceptable. Some people may choose not to reveal their diverse characteristics for fear of discrimination, persecution, or job termination. Therefore, identifying people with specific traits relevant to EDI within the profession may be undesirable or impossible.

Regardless of knowing the prevalence of these additional diversity traits, it is important to understand that these overlap to create a unique set of world views, traditions, and behaviors for each person and community. Intersectionality is the interconnection of how these traits overlap, combine, and interact (Fig. 7). Each person has a unique identity made up of the combination of these traits, which influences interactions in the healthcare environment between patients, healthcare providers, and the community (Chiropractic Educators Research Forum, 2021).

Sometimes outcomes due to diversity traits are advantageous and other times it is not favorable due to discrimination or disadvantage. Healthcare providers must be aware of the established systemic factors that influence healthcare education and the healthcare system to be as effective. When EDI is considered in education, training, and provision of healthcare, we must reflect upon how we consider diversity. We must ask if we create environments that are equitable and promote inclusion, thus directly impacting the health of people, healthcare providers, and the healthcare system. Therefore, throughout their careers, chiropractors must develop and maintain knowledge, skills, and values for diverse traits of their patients and colleagues and know how to effectively function in a diverse healthcare environment (Hayward et al., 2024; Johnson & Green, 2012; Johnson, Killinger, et al., 2012).

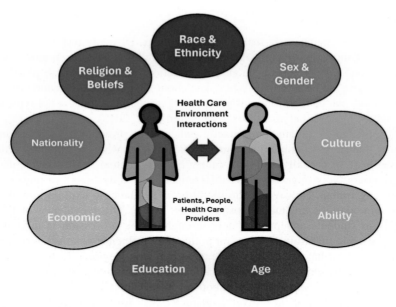

FIG. 7 Diversity traits combine to make an individual's identity, which influence how people interact with one another. *(Modified with permission from Chiropractic Educators Research Forum, presentation Chiropractic Educators Research Forum. (2021). What are diversity? Equity? Inclusion? In:* Presented at preparing for the future: Diversity in chiropractic education: Chiropractic Educators Research Forum (CERF), December 4, 2021. *https://youtu.be/9ZeeAt6SG38.)*

Chiropractic present and future solutions and recommendations

The importance of EDI

It is generally assumed that EDI values are embraced by all healthcare professions. Accordingly, the global chiropractic workforce must fully represent the patients that chiropractors serve. Chiropractic has been credited as being an empathetic and listening profession. However, biases and lack of understanding of people who are different from the provider can be a barrier to effective and high-quality care. It is possible that some people may not recognize the immediate value as EDI relates to chiropractic. Therefore, the following thoughts are offered.

As the world's communities become more diverse, it is important to reflect this in the demographics of the chiropractic profession. The ability to build meaningful therapeutic relationships with patients is an important factor influencing the success of healthcare providers. As such, a provider's understanding and respect of a patient's unique background and culture may help build trust and confidence in the care provided, potentially leading to improved therapeutic outcomes (Yong-Hing & Khosa, 2023).

Chiropractors must be clinically skilled and ready to treat a diverse patient base. This can be accomplished by having a diverse student body, improving skills of awareness of biases, and learning to be active in promoting allyship to those with varied backgrounds or to those who experience inequity. Since educational institutions shape the profession for the next generation, they must promote and implement strategies to meet the future needs of changing patient populations (Ding et al., 2023).

Chiropractors holistically approach patients and embrace the biopsychosocial components of healthcare. To be a holistic healthcare provider, there must be an understanding of the patients being served. Understanding and appreciating the cultural differences of patients can lead to more effective communication and treatment plans that consider patients' unique values, needs, and circumstances. Considering patients' needs is one third of the evidence-based practice model, and therefore requisite to deliver contemporary care. Thus, diversity and cultural competency are relevant and essential components of evidence-based chiropractic practice. Outcomes tend to improve when patients trust their care provider. This underpins the need for a varied and diverse population of healthcare providers.

Within chiropractic practice, there is an emphasis on patient communication, education, and physical touch. Patients' cultural values influence the way they present with and respond to care for musculoskeletal disorders and pain syndromes. Therefore, it is essential for chiropractors to understand EDI and to build safe and trustworthy provider-patient relationships (Amorin-Woods et al., 2021; Stevens, 2007a, 2007b; Weeks et al., 2015; Wong et al., 2022).

The chiropractic profession's future is contingent on developing well-educated, competent practitioners. A diverse and inclusive profession attracts and retains a more diverse pool of chiropractic professionals. This means that people from different backgrounds, including underrepresented groups, are encouraged to pursue a career in chiropractic, which can help to address workforce shortages and enhance the quality of care provided to patients. Students need mentors or role models to which they may aspire or emulate (Bitar et al., 2022; Llamas et al., 2021). Thus, a diverse faculty and staff can aid in growing a diverse student population. Diversity in students enhances the profession by bringing new ideas and questions to the profession to improve the delivery of chiropractic care and provide cultural variety from which the profession may benefit.

Being culturally agile and fluent with EDI concepts benefits chiropractors by allowing them to:

- understand patients' cultural values related to history, examination, and care
- provide culturally responsive care where everyone feels included and respected, regardless of cultural background, ancestry, or lifestyle
- communicate in a thoughtful and empathic manner

- understand the whole person, who is made up of unique traits
- incorporate the understanding of diverse backgrounds and characteristics into practice
- communicate and collaborate effectively with other healthcare providers
- improve patient outcomes and enhance the understanding of how to best serve their communities

EDI benefits the profession at the community and organization levels. By developing initiatives to hire or train healthcare providers from different races, genders, ethnicities, cultures, and backgrounds, a diversity of ideas, thoughts, and beliefs is brought to the profession. This also facilitates the provision of mentors and role models for students from minority groups. Diversity of the workforce is also directly related to reducing burnout and enhancing the well-being of the healthcare workforce (Young et al., 2023). Community interactions enrich the pool of knowledge and the potential for growth for the profession. The diversity of thoughts and ideas leads to new and innovative breakthroughs in healthcare practices.

From a social justice standpoint, healthcare professions are a pathway to financial security. Encouraging admission practices and financial support for underrepresented groups and students from challenged socioeconomic backgrounds may help support underserved communities long-term. The profession must be more inclusive and accessible to a wide range of patients from different backgrounds, including those from historically marginalized communities. The concept of equity is an ethical principle and is consonant with human rights principles (Braveman, 2003, 2010).

EDI creates an environment where people feel included and respected, regardless of their cultural background, ancestry, or lifestyle. Social justice is an important concept related to healthcare (Braveman et al., 2011). In healthcare, everyone should have access to the healthcare services they need, regardless of their social or economic status. For the chiropractic profession, this means that chiropractors should work to ensure that their services are accessible to everyone, regardless of their background or circumstances. By prioritizing social justice and ensuring that everyone has access to chiropractic services, the profession can improve society's overall health and well-being. This not only benefits individuals but also creates a more just and equitable society and in the end also benefits the profession (Hayward et al., 2024).

Chiropractic past and present

Chiropractic grew from people's desire for a more holistic approach to care in the US during the 1800s. However, while emerging as a profession, chiropractic was marginalized and fought against substantial obstacles to be included and recognized as an option for care (Johnson & Green, 2021a, 2021b, 2021c, 2021d). The tumultuous beginnings of the chiropractic profession, and its fight for inclusion and equity in the healthcare arena, may have contributed to the

- WFC position statement: Core elements of chiropractic health care, health promotion, and public health practices (2022) https://www.wfc.org/website/documents/187_core_elements_of_ chiropractic_health_care_health_promotion_and_public_health_practices. pdf

World Spine Care

- World Spine Care Clinics https://www.worldspinecare.org/clinics

Disclosures

No relevant disclosures and no conflicts of interest.

References

Amorin-Woods, L., Gonzales, H., Amorin-Woods, D., Losco, B., & Skeffington, P. (2021). Online or onsite? Comparison of the relative merit of delivery format of aboriginal cultural-awareness-training to undergraduate chiropractic students. *Journal for Multicultural Education*, *15*(4), 374–394.

Amorin-Woods, L. G., Losco, B. E., & Leach, M. J. (2019). A mixed-method study of chiropractic student clinical immersion placements in nonmetropolitan Western Australia: Influence on student experience, professional attributes, and practice destination. *Journal of Chiropractic Education*, *33*(1), 30–39.

Armstrong, K., Ravenell, K. L., McMurphy, S., & Putt, M. (2007). Racial/ethnic differences in physician distrust in the United States. *American Journal of Public Health*, *97*(7), 1283–1289.

Bakaa, N., Southerst, D., Côté, P., Macedo, L., Carlesso, L. C., MacDermid, J., & Mior, S. (2023). Assessing cultural competency among Canadian chiropractors: A cross-sectional survey of Canadian Chiropractic Association members. *Chiropractic & Manual Therapies*, *31*(1), 1–12.

Baker, R. B., Washington, H. A., Olakanmi, O., Savitt, T. L., Jacobs, E. A., Hoover, E., & Wynia, M. K. (2008). African American physicians and organized medicine, 1846-1968: Origins of a racial divide. *JAMA*, *300*(3), 306–313.

Bastos, J. L., Harnois, C. E., & Paradies, Y. C. (2018). Health care barriers, racism, and intersectionality in Australia. *Social Science & Medicine*, *199*, 209–218.

Bitar, J., Montague, G., & Ilano, L. (2022). *Faculty diversity and student success go hand in hand, so why are university faculties so white?*. Education Trust.

Blakemore, E. (2018). *The first birth control pill used Puerto Rican women as guinea pigs*. https://www.history.com/news/birth-control-pill-history-puerto-rico-enovid: History.

Bonifacino, E., Ufomata, E. O., Farkas, A. H., Turner, R., & Corbelli, J. A. (2021). Mentorship of underrepresented physicians and trainees in academic medicine: A systematic review. *Journal of General Internal Medicine*, *36*, 1023–1034.

Brandt, A. M. (1978). Racism and research: The case of the Tuskegee Syphilis Study. *Hastings Center Report*, 21–29.

Braveman, P. A. (2003). Monitoring equity in health and healthcare: A conceptual framework. *Journal of Health, Population and Nutrition*, 181–192.

Braveman, P. (2010). Social conditions, health equity, and human rights. *Health and Human Rights*, *12*, 31.

Braveman, P. A., Kumanyika, S., Fielding, J., LaVeist, T., Borrell, L. N., Manderscheid, R., & Troutman, A. (2011). Health disparities and health equity: The issue is justice. *American Journal of Public Health*, *101*(S1), S149–S155.

Brown, R. A. (2016). Spinal health: The backbone of Chiropractic's identity. *Journal of Chiropractic Humanities*, *23*(1), 22–28.

Brown, B. T., Bonello, R., Fernandez-Caamano, R., Eaton, S., Graham, P. L., & Green, H. (2014). Consumer characteristics and perceptions of chiropractic and chiropractic services in Australia: Results from a cross-sectional survey. *Journal of Manipulative and Physiological Therapeutics*, *37*(4), 219–229.

Callender, A. (2006). Recruiting underrepresented minorities to chiropractic colleges. *Journal of Chiropractic Education*, *20*(2), 123–127.

Chang, E. S., Simon, M., & Dong, X. (2012). Integrating cultural humility into health care professional education and training. *Advances in Health Sciences Education*, *17*(2), 269–278.

Chiropractic Board of Australia. (2023). *Chiropractic annual summary*. https://www.chiropracticboard.gov.au/About-the-Board/Annual-report.aspx.

Chiropractic Educators Research Forum. (2021). What are diversity? Equity? Inclusion? In *Presented at preparing for the future: Diversity in chiropractic education: Chiropractic Educators Research Forum (CERF), December 4, 2021*. https://youtu.be/9ZeeAt6SG38.

Chiropractic Educators Research Forum. (2022a). Academic integrity for all: Building better professionals: Chiropractic Educators Research Forum (CERF), June 25, 2022. *Journal of Chiropractic Education*, *36*, 199–200.

Chiropractic Educators Research Forum. (2022b). Preparing for the future: Diversity in chiropractic education: Chiropractic Educators Research Forum (CERF), December 4, 2021. *Journal of Chiropractic Education*, *36*(2), 194–198.

Christensen, M. G., Morgan, D. R. D., Fetters, S. R., Sieve, Y. D., & Townsend, P. D. (1994). *Job analysis of chiropractic in Australia and New Zealand*. National Board of Chiropractic Examiners (International Division). https://www.ibce.org/wp-content/uploads/2019/08/Job-Analysis-NZ.pdf.

Christensen, M. G., Morgan, D. R. D., Sieve, Y. D., & Townsend, P. D. (1993). *Job analysis of chiropractic in Canada*. National Board of Chiropractic Examiners (International Division). https://www.ibce.org/wp-content/uploads/2019/08/Job-Analysis-in-Canada.pdf.

Coulter, I. D. (1985). The chiropractic patient: A social profile. *The Journal of the Canadian Chiropractic Association*, *29*(1), 25.

Coulter, I. D., & Shekelle, P. G. (2005). Chiropractic in North America: A descriptive analysis. *Journal of Manipulative and Physiological Therapeutics*, *28*(2), 83–89.

Council on Chiropractic Education. (2024). *CCE Accreditation Standards*. Retrieved January 28, 2024 from https://www.cce-usa.org.

Council on Chiropractic Education Australasia. (2017). *Competency Standards for Chiropractors*. Retrieved January 28, 2024 from https://www.ccea.com.au.

Crowther, E. R. (2014). A comparison of quality and satisfaction experiences of patients attending chiropractic and physician offices in Ontario. *The Journal of the Canadian Chiropractic Association*, *58*(1), 24–38.

Danso, R. (2018). Cultural competence and cultural humility: A critical reflection on key cultural diversity concepts. *Journal of Social Work*, *18*(4), 410–430.

Deltoff, M. N. (2023). From student to doctor: An analysis of chiropractic oaths as an allegory of the transition from academic integrity to ethical professionalism. *Journal of Chiropractic Education*, *37*(2), 171–177.

Deltoff, A., & Deltoff, M. N. (1988). One profession—One oath? A survey of the disparity of the chiropractic oath. *Chiropractic History: The Archives and Journal of the Association for the History of Chiropractic, 8*(1), 21–25.

Ding, J., Yong-Hing, C. J., Patlas, M. N., & Khosa, F. (2023). *Equity, diversity, and inclusion: Calling, career, or chore?. Vol. 74* (pp. 10–11). Los Angeles, CA: Sage Publications.

European Council on Chiropractic Education. (2019). *ECCE Standards.* Retrieved January 28, 2024 from https://www.cce-europe.com.

Federation of Canadian Chiropractic. (2018). *Program Standards for the Doctor of Chiropractic Degree Program Canada.* Retrieved January 28, 2024 from https://chirofed.ca.

Field, J. R., & Newell, D. (2016). Clinical outcomes in a large cohort of musculoskeletal patients undergoing chiropractic care in the United Kingdom: A comparison of self-and National Health Service–referred routes. *Journal of Manipulative and Physiological Therapeutics, 39*(1), 54–62.

Fiscella, K., & Sanders, M. R. (2016). Racial and ethnic disparities in the quality of health care. *Annual Review of Public Health, 37*, 375–394.

Forand, D., Drover, J., Suleman, Z., Symons, B., & Herzog, W. (2004). The forces applied by female and male chiropractors during thoracic spinal manipulation. *Journal of Manipulative and Physiological Therapeutics, 27*(1), 49–56.

Freimuth, V. S., Quinn, S. C., Thomas, S. B., Cole, G., Zook, E., & Duncan, T. (2001). African Americans' views on research and the Tuskegee syphilis study. *Social Science & Medicine, 52*(5), 797–808.

Gamble, V. N. (1993). A legacy of distrust: African Americans and medical research. *American Journal of Preventive Medicine, 9*(6), 35–38.

Garcia, J. A., Paterniti, D. A., Romano, P. S., & Kravitz, R. L. (2003). Patient preferences for physician characteristics in university-based primary care clinics. *Ethnicity & Disease, 13* (2), 259–267.

Gaumer, G. (2006). Factors associated with patient satisfaction with chiropractic care: Survey and review of the literature. *Journal of Manipulative and Physiological Therapeutics, 29*(6), 455–462.

Gemmell, H. A., & Hayes, B. M. (2001). Patient satisfaction with chiropractic physicians in an independent physicians' association. *Journal of Manipulative and Physiological Therapeutics, 24* (9), 556–559.

General Chiropractic Council. (2004). *Consulting the profession: A survey of UK chiropractors.* https://www.nightingale-collaboration.org/images/Consulting_the_Profession_A_Survey_of_UK_Chiropractors_2004.pdf.

General Chiropractic Council. (2021). *Registrant survey 2021.* https://www.gcc-uk.org/assets/publications/GCC_Registrant_Survey_2020_-_main_report_final.pdf.

General Chiropractic Council. (2022). *GCC registrant toolkit equality, diversity, and inclusion.* https://www.gcc-uk.org/assets/downloads/GCC_EDI_Toolkit_FINAL_(Web_edition).pdf.

Glucina, T. T., Krägeloh, C. U., & Farvid, P. (2019). Chiropractors' perspectives on the meaning and assessment of quality of life within their practice in New Zealand: An exploratory qualitative study. *Journal of Manipulative and Physiological Therapeutics, 42*(7), 480–491. https://doi.org/10.1016/j.jmpt.2019.02.010.

Godager, G. (2012). Birds of a feather flock together: A study of doctor–patient matching. *Journal of Health Economics, 31*(1), 296–305.

Green, B. N., & Johnson, C. (2010). Chiropractic and social justice: A view from the perspective of Beauchamp's principles. *Journal of Manipulative and Physiological Therapeutics, 33*(6), 407–411. https://doi.org/10.1016/j.jmpt.2010.07.001.

Haldeman, S., Johnson, C. D., Chou, R., Nordin, M., Cote, P., Hurwitz, E. L., Green, B. N., Cedraschi, C., Acaroglu, E., Kopansky-Giles, D., Ameis, A., Adjei-Kwayisi, A., Ayhan, S., Blyth, F., Borenstein, D., Brady, O., Brooks, P., Camilleri, C., Castellote, J. M., … Yu, H. (2018). The global spine care initiative: Care pathway for people with spine-related concerns. *European Spine Journal*, *27*(Suppl 6), 901–914. https://doi.org/10.1007/s00586-018-5721-y.

Haldeman, S., Nordin, M., Chou, R., Cote, P., Hurwitz, E. L., Johnson, C. D., Randhawa, K., Green, B. N., Kopansky-Giles, D., Acaroglu, E., Ameis, A., Cedraschi, C., Aartun, E., Adjei-Kwayisi, A., Ayhan, S., Aziz, A., Bas, T., Blyth, F., Borenstein, D., … Yuksel, S. (2018). The global spine care initiative: World spine care executive summary on reducing spine-related disability in low- and middle-income communities. *European Spine Journal*, *27*(Suppl 6), 776–785. https://doi.org/10.1007/s00586-018-5722-x.

Hammerich, K. F. (2014). Commentary on a framework for multicultural education. *The Journal of the Canadian Chiropractic Association*, *58*(3), 280.

Hawk, C., Long, C. R., Boulanger, K. T., Morschhauser, E., & Fuhr, A. W. (2000). Chiropractic care for patients aged 55 years and older: Report from a practice-based research program. *Journal of the American Geriatrics Society*, *48*(5), 534–545.

Hayward, W. B., Walker, T., & Amorin-Woods, L. G. (2024). Chiropractors can move beyond the deficit discourse toward a strengths-based approach when working with Australia's first peoples. *Chiropractic Journal of Australia*, *51*(1), 1–16.

Himelfarb, I., Hyland, J., Ouzts, N., Russell, M., Sterling, T., Johnson, C., & Green, B. (2020). *Practice analysis of chiropractic 2020*. Greeley, CO: National Board of Chiropractic Examiners.

Human Rights Campaign. *Sexual orientation and gender identity definitions*. https://www.hrc.org/resources/sexual-orientation-and-gender-identity-terminology-and-definitions.

Humphreys, B. K., Peterson, C. K., Muehlemann, D., & Haueter, P. (2010). Are Swiss chiropractors different than other chiropractors? Results of the job analysis survey 2009. *Journal of Manipulative and Physiological Therapeutics*, *33*(7), 519–535.

Imbos, N., Langworthy, J., Wilson, F., & Regelink, G. (2005). Practice characteristics of chiropractors in the Netherlands. *Clinical Chiropractic*, *8*(1), 7–12.

Johl, G. L., Yelverton, C. J., & Peterson, C. (2017). A survey of the scope of chiropractic practice in South Africa: 2015. *Journal of Manipulative and Physiological Therapeutics*, *40*(7), 517–526.

Johnson, V., Assal, S., Khauv, K., Moosad, D., & Morales, B. (2021). Exploring diverse career paths and recommendations for celebrating chiropractic day 2021: A narrative inquiry. *Journal of Chiropractic Humanities*, *28*, 22–34.

Johnson, C., & Green, B. N. (2009). Public health, wellness, prevention, and health promotion: Considering the role of chiropractic and determinants of health. *Journal of Manipulative and Physiological Therapeutics*, *32*(6), 405–412. https://doi.org/10.1016/j.jmpt.2009.07.001.

Johnson, C. D., & Green, B. N. (2012). Diversity in the chiropractic profession: Preparing for 2050. *Journal of Chiropractic Education*, *26*(1), 1–13. https://doi.org/10.7899/1042-5055-26.1.1.

Johnson, C. D., & Green, B. N. (2021a). Looking back at the lawsuit that transformed the chiropractic profession part 1: Origins of the conflict. *Journal of Chiropractic Education*, *35*(S1), 9–24.

Johnson, C. D., & Green, B. N. (2021b). Looking back at the lawsuit that transformed the chiropractic profession part 2: Rise of the American Medical Association. *Journal of Chiropractic Education*, *35*(S1), 25–44.

Johnson, C. D., & Green, B. N. (2021c). Looking back at the lawsuit that transformed the chiropractic profession part 3: Chiropractic growth. *Journal of Chiropractic Education*, *35*(S1), 45–54.

Johnson, C. D., & Green, B. N. (2021d). Looking back at the lawsuit that transformed the chiropractic profession part 8: Judgment impact. *Journal of Chiropractic Education*, *35*(S1), 117–131.

Johnson, C. D., Green, B. N., Agaoglu, M., Amorin-Woods, L., Brown, R., Byfield, D., Clum, G. W., Crespo, W., Da Silva, K. L., & Dane, D. (2023). Chiropractic day 2023: A report and qualitative analysis of how thought leaders celebrate the present and envision the future of chiropractic. *Journal of Chiropractic Humanities, 30*, 23–45.

Johnson, C. D., Green, B. N., Brown, R. A., Facchinato, A., Foster, S. A., Kaeser, M. A., Swenson, R. L., & Tunning, M. J. (2022). A brief review of chiropractic educational programs and recommendations for celebrating education on chiropractic day. *Journal of Chiropractic Humanities, 29*, 44–54.

Johnson, C. D., Haldeman, S., Chou, R., Nordin, M., Green, B. N., Cote, P., Hurwitz, E. L., Kopansky-Giles, D., Acaroglu, E., Cedraschi, C., Ameis, A., Randhawa, K., Aartun, E., Adjei-Kwayisi, A., Ayhan, S., Aziz, A., Bas, T., Blyth, F., Borenstein, D., … Yuksel, S. (2018). The global spine care initiative: Model of care and implementation. *European Spine Journal, 27*(Suppl 6), 925–945. https://doi.org/10.1007/s00586-018-5720-z.

Johnson, C., Killinger, L. Z., Christensen, M. G., Hyland, J. K., Mrozek, J. P., Zuker, R. F., Kizhakkeveettil, A., Perle, S. M., & Oyelowo, T. (2012). Multiple views to address diversity issues: An initial dialog to advance the chiropractic profession. *Journal of Chiropractic Humanities, 19*(1), 1–11. https://doi.org/10.1016/j.echu.2012.10.003.

Johnson, C., Rubinstein, S. M., Cote, P., Hestbaek, L., Injeyan, H. S., Puhl, A., Green, B., Napuli, J. G., Dunn, A. S., Dougherty, P., Killinger, L. Z., Page, S. A., Stites, J. S., Ramcharan, M., Leach, R. A., Byrd, L. D., Redwood, D., & Kopansky-Giles, D. R. (2012). Chiropractic care and public health: Answering difficult questions about safety, care through the lifespan, and community action. *Journal of Manipulative and Physiological Therapeutics, 35*(7), 493–513. https://doi.org/10.1016/j.jmpt.2012.09.001.

Jones, J. M. (2022). *LGBT identification in U.S. ticks up to 7.1%.* Gallup. Retrieved January 15, 2024 from https://news.gallup.com/poll/389792/lgbt-identification-ticks-up.aspx.

Joshi, A., & Ray, S. L. (2019). Facilitators and barriers to education for chiropractic students with visual impairment. *Journal of Chiropractic Education, 34*(2), 116–124.

Khan, J. A., Battaglia, P. J., & Gliedt, J. A. (2023). A narrative review of social determinants of health education in health professional programs and potential pathways for integration into doctor of chiropractic programs. *The Journal of the Canadian Chiropractic Association, 67*(1), 19.

Kopansky-Giles, D., Vernon, H., Steiman, I., Tibbles, A., Decina, P., Goldin, J., & Kelly, M. (2007). Collaborative community-based teaching clinics at the Canadian Memorial Chiropractic College: Addressing the needs of local poor communities. *Journal of Manipulative and Physiological Therapeutics, 30*(8), 558–565.

Lady, S. D., & Burnham, K. D. (2019). Sexual orientation and gender identity in patients: How to navigate terminology in patient care. *Journal of Chiropractic Humanities, 26*, 53–59.

Lekas, H.-M., Pahl, K., & Fuller Lewis, C. (2020). Rethinking cultural competence: Shifting to cultural humility. *Health Services Insights, 13*, 1178632920970580.

Llamas, J. D., Nguyen, K., & Tran, A. G. (2021). The case for greater faculty diversity: Examining the educational impacts of student-faculty racial/ethnic match. *Race Ethnicity and Education, 24*(3), 375–391.

Machovec, C. (2023). *Working women: Data from the past, present and future.* United States Department of Labor. Retrieved March 9, 2024 from https://blog.dol.gov/2023/03/15/working-women-data-from-the-past-present-and-future.

MacMillan, A., Hohenschurz-Schmidt, D., Migliarini, V., & Draper-Rodi, J. (2022). Discrimination, bullying or harassment in undergraduate education in the osteopathic, chiropractic and physiotherapy professions: A systematic review with critical interpretive synthesis. *International Journal of Educational Research Open, 3*, 100105.

Maiers, M. J., Foshee, W. K., & Dunlap, H. H. (2017). Culturally sensitive chiropractic care of the transgender community: A narrative review of the literature. *Journal of Chiropractic Humanities, 24*(1), 24–30.

Marzbanrad, A., Niaghi, F., Tiwana, S., Siddiqi, J., Ding, J., Tanvir, I., Khosa, F., & Tanvir, I., Sr. (2024). Advancing diversity in microbiology: A 55-year retrospective analysis. *Cureus, 16*(1).

Moore, C., Coates, E., Watson, A. R., de Heer, R., McLeod, A., & Prudhomme, A. (2023). "It's important to work with people that look like me": Black patients' preferences for patient-provider race concordance. *Journal of Racial and Ethnic Health Disparities, 10*(5), 2552–2564.

Moscou, K., Bhagaloo, A., Onilude, Y., Zaman, I., & Said, A. (2024). Broken promises: Racism and access to medicines in Canada. *Journal of Racial and Ethnic Health Disparities, 11*, 1182–1198.

National Center for Chronic Disease Prevention and Health Promotion, Office on Smoking and Health; Division of Nutrition, Physical Activity, and Obesity. (2011). *Developing an effective evaluation plan: Setting the course for effective program evaluation.* Atlanta, GA: Centers for Disease Control and Prevention. https://stacks.cdc.gov/view/cdc/24531.

Nelson, L., Pollard, H., Ames, R., Jarosz, B., Garbutt, P., & Da Costa, C. (2021). A descriptive study of sports chiropractors with an International Chiropractic Sport Science Practitioner qualification: A cross-sectional survey. *Chiropractic & Manual Therapies, 29*, 1–6.

Office for National Statistics. (2023). *Sexual orientation, UK: 2021 and 2022.* Retrieved January 15, 2024 from https://www.ons.gov.uk/peoplepopulationandcommunity/culturalidentity/sexuality/bulletins/sexualidentityuk/2021and2022#:~:text=1.-,Main%20points,increase%20from%202.1%25%20in%202017.

Ojo-Aromokudu, O., Suffel, A., Bell, S., & Mounier-Jack, S. (2023). Views and experiences of primary care among black communities in the United Kingdom: A qualitative systematic review. *Ethnicity & Health*, 1–20.

Outerbridge, G., Eberspaecher, S., & Haldeman, S. (2017). World Spine Care: Providing sustainable, integrated, evidence-based spine care in underserved communities around the world. *The Journal of the Canadian Chiropractic Association, 61*(3), 196.

Palmgren, P. J., Chandratilake, M., Nilsson, G. H., & Laksov, K. B. (2013). Is there a chilly climate? An educational environmental mixed method study in a chiropractic training institution. *Journal of Chiropractic Education, 27*(1), 11–20.

Pinn, V. W. (2003). Sex and gender factors in medical studies: Implications for health and clinical practice. *JAMA, 289*(4), 397–400.

Pollentier, A., & Langworthy, J. M. (2007). The scope of chiropractic practice: A survey of chiropractors in the UK. *Clinical Chiropractic, 10*(3), 147–155.

Puhl, A. A., Reinhart, C. J., & Injeyan, H. S. (2015). Diagnostic and treatment methods used by chiropractors: A random sample survey of Canada's English-speaking provinces. *The Journal of the Canadian Chiropractic Association, 59*(3), 279–287.

Renzaho, A., Polonsky, M., McQuilten, Z., & Waters, N. (2013). Demographic and socio-cultural correlates of medical mistrust in two Australian states: Victoria and South Australia. *Health & Place, 24*, 216–224.

Rowell, R. M., & Polipnick, J. (2008). A pilot mixed methods study of patient satisfaction with chiropractic care for back pain. *Journal of Manipulative and Physiological Therapeutics, 31*(8), 602–610.

Sawyer, C., & Kassak, K. (1993). Patient satisfaction with chiropractic care. *Journal of Manipulative and Physiological Therapeutics, 16*(1), 25–32.

Scheffler, R., Cometto, G., Tulenko, K., Bruckner, T., Liu, J., Keuffel, E. L., Preker, A., Stilwell, B., Brasileiro, J., & Campbell, J. (2016). *Health workforce requirements for universal health coverage and the Sustainable Development Goals—Background paper N.1 to the WHO Global Strategy on Human Resources for Health: Workforce 2030.* World Health Organization, issue World Health Organization.

Shamoo, A. E. (2023). Unethical medical treatment and research in US territories. *Accountability in Research, 30*(7), 516–529.

Shippee, T. P., Schafer, M. H., & Ferraro, K. F. (2012). Beyond the barriers: Racial discrimination and use of complementary and alternative medicine among black Americans. *Social Science & Medicine, 74*(8), 1155–1162.

Simpson, J. K., Losco, B., & Young, K. J. (2010). Development of the Murdoch chiropractic graduate pledge. *Journal of Chiropractic Education, 24*(2), 175–186.

Southerst, D., Bakaa, N., Côté, P., Macedo, L., Carlesso, L., MacDermid, J., & Mior, S. (2022). Diversity of the chiropractic profession in Canada: A cross-sectional survey of Canadian Chiropractic Association members. *Chiropractic & Manual Therapies, 30*(1), 1–11.

Statistics Canada. (2024). *Canada at a Glance, 2022 LGBTQ2+ people.* Retrieved January 15, 2024 from https://www150.statcan.gc.ca/n1/pub/12-581-x/2022001/sec6-eng.htm.

Stevens, G. L. (2007a). Behavioral and access barriers to seeking chiropractic care: A study of 3 New York clinics. *Journal of Manipulative and Physiological Therapeutics, 30*(8), 566–572.

Stevens, G. L. (2007b). Demographic and referral analysis of a free chiropractic clinic servicing ethnic minorities in the Buffalo, NY area. *Journal of Manipulative and Physiological Therapeutics, 30*(8), 573–577.

Stochkendahl, M. J., Rezai, M., Torres, P., Sutton, D., Tuchin, P., Brown, R., & Côté, P. (2019). The chiropractic workforce: A global review. *Chiropractic & Manual Therapies, 27*, 1–9.

Stronach, M., O'Shea, M., & Maxwell, H. (2023). 'You can't be what you can't see': Indigenous Australian sportswomen as powerful role models. *Sport in Society, 26*(6), 970–984.

Torgrimson, B. N., & Minson, C. T. (2005). Sex and gender: What is the difference? *Journal of Applied Physiology, 99*, 785–787.

Vindigni, D., Polus, B., Edgecombe, G., van Rotterdam, J., Turner, N., Spencer, L., Irvine, G., & Walsh, M. (2009). Bringing chiropractic to aboriginal communities: The Durri model. *Chiropractic Journal of Australia, 39*(2), 80–83.

Vollenweider, R., Peterson, C. K., & Humphreys, B. K. (2017). Differences in practice characteristics between male and female chiropractors in Switzerland. *Journal of Manipulative and Physiological Therapeutics, 40*(6), 434–440.

Weeks, W. B., Goertz, C. M., Meeker, W. C., & Marchiori, D. M. (2015). Public perceptions of doctors of chiropractic: Results of a national survey and examination of variation according to respondents' likelihood to use chiropractic, experience with chiropractic, and chiropractic supply in local health care markets. *Journal of Manipulative and Physiological Therapeutics, 38*(8), 533–544.

Weigel, P. A., Hockenberry, J. M., & Wolinsky, F. D. (2014). Chiropractic use in the Medicare population: Prevalence, patterns, and associations with 1-year changes in health and satisfaction with care. *Journal of Manipulative and Physiological Therapeutics, 37*(8), 542–551.

Wiese, G. S. C. (2003). *Choices, challenges, and leaps of faith: A qualitative study of sixteen African-American chiropractors' experiences in chiropractic education.* The University of Iowa.

Wilson, T., Temple, J., Lyons, A., & Shalley, F. (2020). What is the size of Australia's sexual minority population? *BMC Research Notes, 13*, 1–6.

Wong, Y. K., Low, K. L., & Pooke, T. G. (2022). Factors associated with dimensions of patients' trust in chiropractic doctors in the International Medical University Healthcare Chiropractic Center: An exploratory study. *Journal of Chiropractic Medicine*, *21*(2), 83–96.

World Federation of Chiropractic. (2001). *Definition of Chiropractic*. Retrieved January 15, 2024 from https://www.wfc.org/website/index.php?option=com_content&view=article&id=90& Itemid=110&lang=en.

World Federation of Chiropractic. (2020). *Equity, diversity, and inclusion policy (September 2020)*. Retrieved February 20, 2024 from https://www.wfc.org/website/images/wfc/Policies/Equity_ Diversity_Inclusion_policy_2020.pdf.

World Health Organization. (2023). *The programme for gender equality, human rights & health equity*. Retrieved March 11, 2024 from https://www.who.int/teams/gender-equity-and-human-rights/about.

Xu, R. Y., Sood, N., Ding, J., Khan, N., Tiwana, S., Siddiqi, J., … Khosa, F. (2024). End of affirmative action—Who is impacted most? Analysis of race and sex among US internal medicine faculty. *Journal of General Internal Medicine*, *39*(9), 1556–1566.

Yelverton, C., Zietsman, A., Johl, G., & Peterson, C. (2022). Differences in practice characteristics between male and female chiropractors in South Africa: Secondary analysis of data from scope of practice survey 2015. *Journal of Manipulative and Physiological Therapeutics*, *45*(1), 90–96.

Yong-Hing, C. J., & Khosa, F. (2023). Provision of culturally competent healthcare to address healthcare disparities. *Canadian Association of Radiologists Journal*, *74*(3), 483–484.

Young, K. J. (2015). Overcoming barriers to diversity in chiropractic patient and practitioner populations: A commentary. *Journal of Cultural Diversity*, *22*(3).

Young, P. J., Kagetsu, N. J., Tomblinson, C. M., Snyder, E. J., Church, A. L., Mercado, C. L., Perez-Carrillo, G. J. G., Jha, P., Guerrero-Calderon, J. D., & Jaswal, S. (2023). The intersection of diversity and well-being. *Academic Radiology*, *30*(9), 2031–2036.

Zhu, K., Das, P., Karimuddin, A., Tiwana, S., Siddiqi, J., & Khosa, F. (2021). Equity, diversity, and inclusion in academic American surgery faculty: An elusive dream. *The Journal of Surgical Research*, *258*, 179–186. https://doi.org/10.1016/j.jss.2020.08.069.

of Canada, consideration must be given to the procedural barriers to practice for internationally trained dentists. Given, the racialized makeup of these immigrants and the colonial history of Canada, questions arise regarding lack of EDI. These aspirants must go through an equivalency process through the National Dental Examining Board of Canada (NDEB). This process has been consistently criticized as long, arduous, and expensive so it may take aspirants 3–5 years or longer to complete (Rowe, 2023). The years spent preparing, taking, and clearing the qualifying examinations are seen as a professional/career "break," therefore putting the candidate at an additional unearned disadvantage.

A review of the global gender make-up of the dental workforce highlighted that in 2016, there were a substantial number of women dental school graduates across Europe (Tiwari et al., 2019). Women made up at least 70% of the graduating classes in 10 nations including but not limited to Belgium, Denmark, Estonia, Lithuania, Norway, Poland, Romania, and Slovenia. While they made up 55% of dental graduates in France and 56% in the United Kingdom, respectively (Kravitz et al., 2016). In other parts of the world, such as Japan, women made up 41% of all dentistry graduates (Japan Dental Association, 2014). Over 50% of registered dentists in some Latin American nations, including Brazil and Chile, were women (Counsel of Federal Odontologia, 2018; Guíñez et al., 2018). Women made up 33.2% and 36.8% of registered dentists in Australia and New Zealand in 2012, respectively (Australian Institute of Health and Welfare, 2014). For Asian and African nations, statistics on the representation of women in dentistry is limited. For instance, 53% of dentists who were registered in Cameroon were women (Agbor et al., 2018). Women made up 35% of the registered dentists in Rwanda and 40% in the Democratic Republic of the Congo in 2014, respectively (World Dental Federation, 2015). The percentages were 32% in South Korea and Hong Kong SAR (World Dental Federation, 2015). In Pakistan 52% of registered dental surgeons were women, according to the Pakistan Medical and Dental Council (2018). Notably, the most recent data from the Dental Council of India (2018) showed that women made up 70% of all registered dentists in India. Despite this expansive presence in the dental workforce, the representation of women in academic faculty is limited, with a higher percentage of men across senior academic ranks and leadership positions. Taking two distinct universities as case study examples, University of Nairobi in Kenya and University of Copenhagen in Denmark. At the University of Nairobi, women made up 36% of the lecturers and 41% of the entire academic faculty. Women, however, were underrepresented in positions of influence and decision making, such as full professorships or department heads. At the University of Copenhagen in Denmark, women made up 62% of the faculty as of the most recent data. Unfortunately, just 22% of them were full professors, which shows a lack of representation in higher academic ranks (Tiwari et al., 2019).

Critically analyzing the United Kingdom (UK) dental workforce pipeline, we can gauge the underrepresentation of racialized minorities from dental education to practice. The pooling of different racial and ethnic identities in the census makes it difficult to assess the extent of marginalization in dental academia and practice for each social identity. However, inferences have been made with existing census and survey data. Among the entire dental workforce in the UK, 52% identified as non-White in 2015. In contrast, 81.4% of clinical academic appointments were White with a meager 18.6% of other categories including 12.4% Asian, 1.7% Chinese, 1.1% Black, 1.1% Mixed, and 2.3% others (Advance HE, 2015). Improvements in this regard have been noticed with an increase to 22% of non-White clinical academics in 2018. This may be reflective of systemic barriers and lack of opportunities in academic settings. Similarly, White students have been twice as more likely to graduate with honors than Black students (Mountford-Zimdars et al., 2015). Meanwhile, despite these academic barriers, Black students have been equally as likely as White students to gain specialist accreditation. However, Asian and those with mixed heritage are least likely to pursue specialization (Lala et al., 2021).

According to the 2018–19 census, the US dental school faculty members were predominantly White (61.7%). This was a 10% drop from the prior census conducted in 2011–12. The percentages of HURE (Hispanic/Latinx, non-Hispanic Black/African American, non-Hispanic American Indian/Alaska Native, non-Hispanic Native Hawaiian/Other Pacific Islander) full-time and part-time dental school faculty declined from 13.4% to 11.1% over the same period. Faculty of color also experienced a slight decline from 23.9% to 23.3% (Smith et al., 2022). For women specifically, without the overlaying intersect of race/ethnicity, 37% of the full-time dental faculty positions were occupied by women (American Dental Education Association, 2015). Further findings from 2016 show that 32% of primary administrative positions in dental education were held by women, and at the highest leadership level in dental education, there were 13/66 (20%) women dental school deans, representing a disproportionate number of women academics (American Dental Education Association, 2016). The UK showed a similar trend with an increase from 4.8% in racial diversity in the early 2000s to 6.7% in 2014 among the country's dental academics (Watson et al., 2018).

The American Dental Education Association (ADEA) has committed to the development of an inclusive environment in which faculty, staff, and administrators help shape and promote priorities representing EDI within dental education. ADEA has organized and/or facilitated workshops, seminars, and symposiums to address the inequalities in accessing dental education by various gender identities, ethnicities, and backgrounds. It has focused on leveraging in-group identity to eliminate bias and foster humanistic and inclusive environments in support of dental school and allied dental program community stakeholders (ADEA, 2019). While these are admirable endeavors and do in fact aid in combating existing patriarchal and colonial organizational cultures and attitudes. They do little to address the gaps between policy formulation and its actual implementation in dental education institutions. The translation of policy into practice often requires adequate resources, including funding, protected time, safe environment, recognition, protection from potential retribution, personnel, infrastructure, and technology. Sometimes the intention to improve is present, but due to Institutional inertia, lack of buy-in from stakeholders, and resistance from employees who benefit from the status quo progress remains elusive. EDI policies are complex and often require coordination and collaboration among multiple stakeholders, departments, or organizations. Hence, a lack of clear communication, coordination mechanisms, and collaboration frameworks can result in disjointed efforts and hinder effectiveness.

Addressing these intractable issues requires proactive planning, stakeholder engagement, resource allocation, clear communication, capacity building, and continuous monitoring and evaluation for course correction. Additionally, a supportive institutional culture and leadership commitment are vital to successfully convert policy into practice. Within the leadership, there should be a clear understanding of the goals and objectives at each level of the hierarchy and knowledge of best practices or principles to create an inclusive supporting environment. Fig. 1 shows the multifaceted dimensions of EDI policies across varying levels of leadership, their interactions, and the goals to achieve.

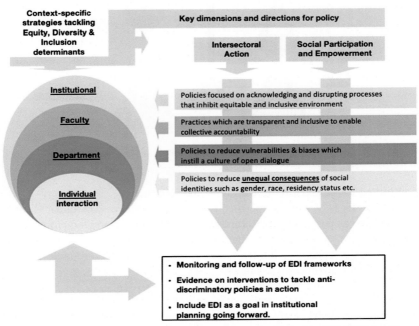

FIG. 1   Dimensions of policy in equity, diversity, and inclusion determinants. *(Modified and based on the World Health Organization Social Determinant of Health model (Solar, O., & Irwin, A. (2018). A conceptual framework for action on the intersectional approaches to EDI initiatives.)*

The advantages of EDI have been observed to encompass both concrete and intangible aspects. Institutions that have prioritized EDI have seen positive changes in employee satisfaction, better retention and engagement, and enhanced success and profitability (Young et al., 2023). Research has shown that diversity among working groups improves problem-solving skills (Anand & Winters, 2008; Marder, 2017). The understanding of a diverse healthcare workforce, particularly in dentistry, is a relatively newer concept. It stands to reason that the advantages of diversity in other sectors would also be true for the dental healthcare sector. It is also crucial to keep in mind that addressing health is not simply a clinical solution but requires a broader social lens, which depends on varying inclusive perspectives from patients, caregivers, policymakers, community leaders, and researchers **(Image source: Elsevier, shutterstock_1919104061)**.

Potential barriers

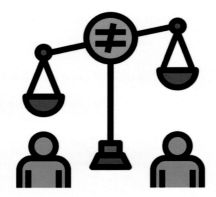

(Source: Elsevier, shutterstock_1899553327.)

1. **Limited Representation in Curricula and Research**
   The lack of diverse perspectives and experiences in curricula and research can perpetuate biases and reinforce existing power structures. Limited representation can contribute to the marginalization of certain knowledge and perspectives, hindering the development of a more inclusive and nurturing academic environment.
2. **Lack of Diversity in Leadership Positions**
   The lack of diverse representation within leadership positions, such as department chairs, deans, and university administrators, can hinder progress toward EDI. Lack of diverse perspectives in decision-making can result in policies and practices that do not adequately address the needs and experiences of underrepresented groups.

Baah, F. O., Teitelman, A. M., & Riegel, B. (2019). Marginalization: Conceptualizing patient vulnerabilities in the framework of social determinants of health—An integrative review. *Nursing Inquiry, 26*(1), e12268. https://doi.org/10.1111/nin.12268.

Bollinger, L. C. (2021). *Our accelerated faculty diversity efforts.* Columbia University Office of the President. [Internet]. Retrieved from https://president.columbia.edu/news/our-accelerated-faculty-diversity-efforts.

Cain, L., Brady, M., Inglehart, M. R., & Istrate, E. C. (2022). Faculty diversity, equity, and inclusion in academic dentistry: Revisiting the past and analyzing the present to create the future. *Journal of Dental Education, 86*(9), 1198–1209.

Caiola, C., Docherty, S., Relf, M., & Barroso, J. (2014). Using an intersectional approach to study the impact of social determinants of health for African American mothers living with HIV. *ANS. Advances in Nursing Science, 37*(4), 287–298. https://doi.org/10.1097/ ANS.0000000000000046.

Canadian Gender Budgeting Act. (2018). *S.C 2018, C. 27, S.314.* https://laws-lois.justice.gc.ca/ PDF/C-17.2.pdf.

Carnes, M., Fine, E., & Sheridan, J. (2019). Promises and pitfalls of diversity statements: Proceed with caution. *Academic Medicine, 94*(1), 20–24. https://doi.org/10.1097/ ACM.0000000000002388.

Counsel of Federal Odontologia. (2018). *Council creates FAQ for clarification of frequently asked questions about edited resolutions.* http://cfo.org.br/website/.

Dental Council of India. (2018). *Gender profile.* http://www.dciindia.org.in/GrpReportIDR.aspx.

Fiala, J. (2017, April 24). AAVMC: Fewer men, more debt in veterinary academia. *VIN News Service.* http://news.vin.com/vin-news.aspx?articleId=44613.

Garg, P. S., Aagaard, E., Amiel, J., Boutin-Foster, C. B., Mann, S., Markell, M., et al. (2021). *Creating action to eliminate racism in medical education: Medical education senior leaders' rapid action team to combat racism in medical education.* AAMC.

Gonzales, M. (2023, June 30). *Supreme court dismantles affirmative action in college admissions.* Society for Human Resource Management. Retrieved July 12, 2023, from https://www.shrm. org/resourcesandtools/legal-and-compliance/employment-law/pages/affirmative-action-supreme-court-cases.aspx.

Guíñez, J., Guajardo, P., Cartes-Velásquez, R., & Campos, V. (2018). The current status of dental education and the dental profession in Chile. *Pesquisa Brasileira em Odontopediatria e Clinica Integrada, 18*(1), e3875.

Japan Dental Association. (2014). *Website in Japanese.* https://www.jda.or.jp/dental_data/pdf/ chapter_04.pdf.

Kravitz, A., Bullock, A., Cowpe, J., & Barnes, M. (2016). *Manual of dental practice 2015* (5.1st ed.). Brussels: Council of European Dentists. https://cedentists.eu/library/eu-manual.html.

Lala, R., Baker, S. R., & Muirhead, V. E. (2021). A critical analysis of underrepresentation of racialised minorities in the UK dental workforce. *Community Dental Health, 38*(2), 142–149. https:// doi.org/10.1922/CDH_IADRLala08.

Magee, M. (2016, March 9). The diversity crisis for education's leading roles. *Education Week, 35*(23), 21. https://www.edweek.org/ew/articles/2016/03/09/why-is-education-leadership-so-white.html.

Marder, A. (2017). *7 studies that prove the value of diversity in the workplace.* Capterra. https:// www.capterra.com/resources/7-studies-that-prove-the-value-of-diversity-in-the-workplace/.

McKay, J. C., & Quiñonez, C. R. (2012). The feminization of dentistry: Implications for the profession. *Journal of the Canadian Dental Association, 78*(1), 7.

Mountford-Zimdars, A., Sabri, D., Moore, J., Sanders, J., Jones, S., & Higham, L. (2015). *Causes of differences in student outcomes (HEFCE)*. London: Higher Education Funding Council for England. https://dera.ioe.ac.uk/23653/1/HEFCE2015_diffout.pdf.

Pakistan Medical and Dental Council. (2018). *Pakistan Medical and Dental Council Statistics*. http://www.pmdc.org.pk/statistics/tabid/103/default.aspx.

Preston, J. (2008). The urgency of postsecondary education for aboriginal peoples. *Canadian Journal of Educational Administration and Policy*, 86, 1–22.

Rowe, B. (2023). *Internationally trained dentists voice frustration with credential process in Canada*. https://toronto.citynews.ca/2023/02/05/international-dentists-credentials-canada-toronto/.

Smith, S. G., Banks, P. B., Istrate, E. C., Davis, A. J., Johnson, K. R., & West, K. P. (2022). Antiracism structures in academic dentistry: Supporting underrepresented racially/ethnically diverse faculty. *Journal of Public Health Dentistry*, *82*(Suppl 1), 103–113. https://doi.org/10.1111/jphd.12509.

Sotto-Santiago, S., Sharp, S., Mac, J., Messmore, N., Haywood, A., Tyson, M., & Yi, V. (2021). Reclaiming the mission of academic medicine: An examination of institutional responses to (anti)racism. *AEM Education and Training*, *5*(S1), S33–S43. https://doi.org/10.1002/aet2.10668.

StatsCan, Govt of Canada. (2022). *Racialized groups*. https://www150.statcan.gc.ca/n1/pub/12-581-x/2022001/sec3-eng.htm.

Tiwari, T., Randall, C. L., Cohen, L., Holtzmann, J., Webster-Cyriaque, J., Ajiboye, S., … D'souza, R. N. (2019). Gender inequalities in the dental workforce: Global perspectives. *Advances in Dental Research*, *30*(3), 60–68.

Tsai, L. L., Ha, J. S., & Yang, S. C. (2021). Asian Americans (AAPI) in academic surgery. *Annals of Surgery*, *274*(6), e628–e629.

UI College of dentistry students issue demands for administration during protest. (2021). Des Moines Register. [Internet]. Retrieved from https://www.desmoinesregister.com/picture-gallery/news/education/university-of-iowa/2021/01/30/action-uiowa-students-protest-discrimination-dental-school/4315123001/ (cited 14 October 2021).

University Affairs. (2020). *Study finds equity, diversity and inclusion policies off to a "strong start" at research-intensive universities*. https://www.universityaffairs.ca/news/news-article/study-finds-equity-diversity-and-inclusion-policies-off-to-a-strong-start-at-research-intensive-universities/ (Accessed 12 May 2023).

Versaci, M. B. (2021). *HPI: Women make up growing percentage of dental workforce*. https://www.ada.org/en/publications/ada-news/2021/march/women-make-up-growing-percentage-of-dental-workforce.

WAGE. (2021, April 14). *Women and gender equality Canada*. Government of Canada. Retrieved April 27, 2023, from https://women-gender-equality.canada.ca/en/gender-based-analysis-plus/resources/action-plan-2016-2020.html.

WAGE (Women and Gender Equality Canada). (2021). *Government of Canada announces recipients of $100-million Feminist Response and Recovery Fund*. (Accessed 10 April 2023).

Wang, K. (2020). Among other measures, UW must include additional training dedicated to racial inclusivity. Following conclusion to Breonna Taylor investigation, UW should mandate additional racial sensitivity training to faculty, students. *The Badger Herald*. [Internet]. Retrieved from https://badgerherald.com/opinion/2020/09/29/among-other-measures-uw-must-include-additional-training-dedicated-to-racial-inclusivity/ (cited 14 October 2021).

Watson, N., Tang, P., & Knight, E. (2018). *Survey of dental clinical academic staffing levels*. London: Dental Schools Council. https://www.dentalschoolscouncil.ac.uk/wp-content/uploads/2018/08/clinical-academic-survey-dental-2018.pdf.

Wei, W., et al. (2023). Organizational leadership gender differences in medical schools and affiliated universities. *Journal of Women's Health, 33*(5). https://doi.org/10.1089/jwh.2023.0326. In this issue.

White House, Gov of USA. (2021). *Executive order on diversity, equity, inclusion, and accessibility in the federal workforce.* https://www.whitehouse.gov/briefing-room/presidential-actions/2021/06/25/executive-order-on-diversity-equity-inclusion-and-accessibility-in-the-federal-workforce/.

World Dental Federation. (2015). The challenge of oral disease—A call for global action. In *The oral health atlas* (2nd ed.). Geneva: FDI World Dental Federation. https://www.fdiworlddental.org/sites/default/files/media/documents/complete_oh_atlas.pdf.

Yong-Hing, C. J., & Khosa, F. (2023). Provision of culturally competent healthcare to address healthcare disparities. *Canadian Association of Radiologists Journal = Journal l'Association Canadienne des Radiologistes, 74*(3), 483–484. https://doi.org/10.1177/08465371231154231.

Young, P. J., Kagetsu, N. J., Tomblinson, C. M., Snyder, E. J., Church, A. L., Mercado, C. L., Guzman Perez-Carrillo, G. J., Jha, P., Guerrero-Calderon, J. D., Jaswal, S., Khosa, F., & Deitte, L. A. (2023). The intersection of diversity and well-being. *Academic Radiology, 30*(9), 2031–2036. https://doi.org/10.1016/j.acra.2023.01.028.

From this confluence of societal developments, home economics as an area of learning and practice came to the fore and eventually became a discipline of study in colleges and universities.

Women at the forefront of the home economics movement were mainly racialized as white and from upper-middle-class households, though they diverged in terms of their political/societal viewpoints. The proponents advocated for women's education to "prepare girls for their future roles as homemakers" and considered domestic work as requiring "specialized training" and deserving of prestige (Brady, 2017). This group disparaged women's entry into paid professional employment outside of the home, rather advocating that women would be fulfilled through their domestic duties. The opposition group sought to advance women's roles in public life. They advocated for scientific training for women in order to position women to help address the issues central to the times, a position inconsistent with the largely male-dominated biological sciences (Gingras et al., 2014). In this way, the opposition contended, women could enter paid professional employment. Home economics as a discipline was officially adopted in 1899 at the first of a series of conferences held in Lake Placid, NY (Brady, 2017).

Sometime after the realization of home economics as a discipline of study, it evolved to become specialized. Whereas once the discipline centered on nutrition/dietetics, child development, personal finance, and clothing/textiles, it became less cohesive, and areas within the discipline became less valued. Additionally, home economics was influenced by late 19th and early 20th century notions of race, class, and gender. In this way, colonialist, racist, classist, and sexist ideas informed home economics education and practice. For example, African American women activists saw domestic science education as a way to promote economic opportunity and racial uplift (Scott, 2009), though they had to confront deeply entrenched stereotypes of Black women as domestics and cooks. Black domestic science programs expanded during the WWI era, though racist sentiments regarding the status of "cooks" led white dietitians to distance themselves by labeling themselves "scientific cooks." At that time, dietitians also complained of their "lack of professional autonomy, control and status" (Scott, 2009, p. 16); concerns that are still being voiced today.

The racist discourse also played out in the eugenics movement,[b] where studies of human nutrition, US public policy, and the marketing of "normative" standards of middle-class, Protestant, white womanhood (Scott, 2009) were amplified. At this time, dietitians developed their own forms of knowledge

---

b. Indeed, one of the "founders" of the American Dietetic Association (now the Academy of Nutrition and Dietetics), Lenna Frances Cooper, through her work as Dean of the Battle Creek Sanitarium School of Home Economics (Battle Creek, MI) with Dr. John and Mrs. Ella Kellogg, would likely have participated in the eugenics movement as the sanitarium was the repeated site of the National Conference on "Race Betterment." Similarly, one of the presidents of the American Dietetic Association (1920–22), Mary De Garmo Bryan, was the daughter of Mary Eliose De Garmo, an American educator and "clubwoman" known as the originator of the "Better Babies Contests" held throughout the southern US to promote infant health and eugenics (Scott, 2009).

and systems based upon that knowledge, and the profession distanced itself from groups in higher education and the military (Scott, 2009). Professionalism within the white, Protestant, middle-class woman medical hierarchy relied upon the "exclusion and subordination of ethnic women and women of color" (Scott, 2009, p. 14). As a means of reinforcing this prejudice, dietitians crafted journal articles that "scientifically" linked dietary habits and deficiency diseases with specific racial, ethnic, religious, and regional groups (Scott, 2009). Additionally, they elevated themselves above nurse's aides, cooks, and kitchen employees through the assertion of a presumed Anglo-American cultural superiority (Scott, 2009).

With the dietetics profession emerging from either nursing training programs or home economics, there arose confusion within the nursing profession over roles and authority. Dietitians responded by purposefully situating themselves between the patient and the physician (Scott, 2009). They catered to male physicians and claimed an alliance with the medical profession; this alliance further distanced dietetics from home economics (Scott, 2009). As such, didactic programs in dietetics incorporated more and more "hard sciences" into their curricula. These changes remain today where those studying to become RD(N)s must successfully complete coursework in biology, chemistry, biochemistry, anatomy, and physiology, to name a few.

At present, dietetics in the US is hesitant to acknowledge its history in terms of its racist roots. It is rare that the history of the profession is discussed as part of the didactic or experiential components of dietetics education, let alone the history of racism inherent in the profession. An example of this reticence by the Academy to acknowledge its racist roots is demonstrated in the recent (2021) renaming of the Lenna Frances Cooper Memorial Lecture Award to the Distinguished Lecture Award. This change was prompted by some members of the Academy notifying the then President Dr. Kevin Saur of the racist actions associated with Ms. Cooper. After conducting research into this claim, the Academy voted to make the name change, announcing it in an Academy alert that was distributed to its membership over the 2021 US Thanksgiving holiday, without mentioning the real reason why the award was being renamed.

Relevant history of dietetics in Canada

Dietetics in Canada, like other health professions, has benefited from and developed through a history of colonization and participation in the active subjugation of Indigenous peoples and their knowledge systems. Between the 1900s and 1940s, around the time that the profession as a discipline was evolving, Indigenous peoples globally were being subjected without consent to clinical and anthropological research. Some of this research was referred to as "salvage research" and was predicated on Indigenous peoples as a group being destined to go extinct; contentions that were based upon theories of evolution and informing eugenics movements (Gruber, 1970; Wilson, 2008). Because of this, of the (Nutrition) Dietetics Technician Registered [(N)DTR] credential with a transition of those professionals from worksites such as the hospital setting providing care and supporting the RD(N) in nutrition assessment and care plan formation/implementation to "local communities." Given that (N)DTR credentialed professionals, unlike RD(N) credentialed professionals, are more diverse in terms of race/ethnicity, this move would reduce diversity in the profession. Another change proposed (which has since been enacted) is for the educational preparation of further entry-level RD(N)s to a minimum of a graduate degree. Despite years of pushback from Academy membership who fear the detrimental effect this "Master's degree mandate" will have on diversity in the discipline, as well as the lack of evidence to support CDR's contention that the graduate degree mandate will improve the quality of care provided by RD(N)s, patient safety, and salaries, the CDR imposed the mandate which took effect January 1, 2024.

Many individuals have spoken out against CDRs Master's degree mandate, including a professor at New York University who was quoted in the New York Times article as saying the mandate is "unconscionable" and "an even greater barrier to people of color [to enter] our profession" (NYT). The 2017 Visioning Report (Kicklighter et al., 2017) called for increasing diversity within the profession to (particularly) improve healthcare for underrepresented communities and populations. This recommendation cannot, nor will not, be realized with the increased educational requirements at the graduate level which will most burden individuals without the economic means to complete a graduate degree.

Review of longitudinal trends in DEI in dietetics in the United States and Canada

In 1987, the American Dietetic Association (as it was called then) created its first effort to address the lack of diversity in the discipline. At that time, the Affirmative Action Committee was established. Demographic data of dietetics professionals [including race/ethnicity and "sex" (binary)] were first reported to the membership in 1990 in a series of reports on the membership database of the American Dietetic Association (the Academy). It is unclear if these data were collected in any systematic manner prior to 1990. These 1990 data included demographic information for both retired and active RD(N)s and NDTRs, with NDTR professionals traditionally being a more diverse group. After 1990, demographic data were collected separately for these groups of professionals. The published demographic data are available for the years 1990, 1992, 1993, and 1995 (Bryk & Kornblum, 1991, 1993; Bryk & Soto, 1994, 1997). In 1995, the Academy renamed the Affirmative Action Committee the Diversity Committee. Data were again collected in 1997 (Bryk & Soto, 1999). In 1998, recognizing that effort would be needed to increase diversity in the profession, the Academy indicated a commitment to "increasing diversity in educational preparation of underrepresented groups by five percentage points during a six year period" (Stein, 2012). Interestingly, no demographic data are available via the membership database reports after 1999 (Bryk & Soto, 2001). In addition, there are no data available either during this period or forward from these dates on the number of dietetics professionals or students with disabilities (Baxter et al., 2020).

In 2000, the Academy developed and issued a free "Diversity Toolkit" to members upon request, established a bi-annual $10k Diversity & Inclusion Promotion Grant,[d] and issued a Diversity Philosophy Statement (Stein, 2012). Additionally, it supported the creation of Member Interest Groups, founded the Diversity Action and Diversity Leader Program awards, generated resources on cultural competency, and included multiethnic images in its messaging (Stein, 2012). Although 2001–08 showed overall gains in the ethnic diversity of students enrolled in accredited programs compared to 1993–2000, these increases were made up predominately by Hispanic and Asian students as the percentage of Black-identified students decreased. There was also a decrease in students who identified as male over the same time frame. Overall, the goal to increase the number of dietetics professionals who identify as members of minoritized communities "has not yet been realized" and progress to date has been relatively minimal (Stein, 2012, p. 794).

The year 2004 was the first where demographic data regarding dietetics professionals was made available through the American Dietetic Association's Member Needs Assessment/Satisfaction Study, which is published every four - years. Data presented in the study were obtained from a "representative sample" of dietetics professionals who are members of the Academy (Rogers, 2000, 2005, 2009, 2013, 2017, 2021). Nonmembers were not included.[e] Data published from this survey were reported in percentages, and gender was queried as binary (man, woman, no answer). The 2016 data mirrored the data for Black-identifying members obtained in 2008 (2%), and Hispanic-identified members increased from 3% in 2008 to 4% in 2016.

The Diversity Committee of the Academy was renamed the Diversity and Inclusion Committee in 2018, and in 2020, the committee's name was changed again to the Inclusion, Diversity, Equity and Access Committee (IDEA). The 2020 data published on the CDR website before it was removed showed the percentage of credentialed professionals (not just members of the Academy) who identified as Black at 2.6%. In the 2020 Member Needs Assessment/Satisfaction Survey, this number was 3%. When asked why the CDR data were removed from the website, members were told that the Member Needs Assessment/

---

d. Of note, these grants are applied for by and granted only to Academy members who are appointed as "Diversity Liaisons" to Academy Dietetic Practice Groups (DPGs). Grant awardees are not required to submit outcomes data in terms of their success increasing diversity and inclusion in the profession.

e. Currently, the Academy boasts "over 112,000 members" (RD(N)s, (N)DTRs, and "other professionals". Data regarding how many RD(N)s are not members of the Academy is not available.

Gruber, J. W. (1970). Ethnographic salvage and the shaping of anthropology. *American Anthropologist*, *72*(6), 1289–1299. http://www.jstor.org/stable/672848.

Grundy, S. (2022). *Respectable: Politics and paradox in making the morehouse man*. Univ of California Press.

Health Canada. (2010). *Eating well with Canada's Food Guide - First nations, inuit and métis*. https://www.canada.ca/en/health-canada/services/canada-food-guide/about/history-food-guide/eating-well-canada-food-guide-first-nations-inuit-metis.html. (Accessed 31 July 2024).

Hwalla, N., & Koleilat, M. (2004). Dietetic practice: The past, present and future. *EMHJ-Eastern Mediterranean Health Journal*, *10*(6), 716–730.

Joy, P., & Hawthorne, L. (2023). The Schooling of Peter Pan: Constructs of Gender and Sexual Orientation within Dietetics. *Journal of Critical Dietetics*, *6*(3), 5–29.

Joy, P., & McSweeney-Flaherty, J. M. (2022). Moving dietetics forward with queer pedagogy: A post-structural qualitative study exploring the education and training experiences of Canadian dietitians for LGBTQ care. *Journal of the Academy of Nutrition and Dietetics*, *122*(10), 1876–1884.

Kicklighter, J. R., Cluskey, M. M., Hunter, A. M., Nyland, N. K., & Spear, B. A. (2013). Council on future practice visioning report and consensus agreement for moving forward the continuum of dietetics education, credentialing, and practice. *Journal of the Academy of Nutrition and Dietetics*, *113*(12), 1710–1732.

Kicklighter, J. R., Dorner, B., Hunter, A. M., Kyle, M., Prescott, M. P., Roberts, S., ... Byrne, C. (2017). Visioning report 2017: A preferred path forward for the nutrition and dietetics profession. *Journal of the Academy of Nutrition and Dietetics*, *117*(1), 110–127.

Krisna, P. (December 7, 2020). *Is American dietetics a White-bread world*. NewYorkTimes.com. https://www.nytimes.com/2020/12/07/dining/dietitian-diversity.html (Accessed 15 January 2023).

Mahajan, A., Banerjee, A. T., Ricupero, M., Beales, A., Lac, J., Ajwani, F., ... Pais, V. (2021). Call to action to improve racial diversity in dietetics. *Journal of Critical Dietetics*, *5*(2), 3–9.

Mosby, I. (2013). Administering colonial science: Nutrition research and human biomedical experimentation in Aboriginal communities and residential schools, 1942–1952. *Histoire Sociale/Social History*, *46*(1), 145–172.

Nova Scotia Dietitians Association. *Continuing Competency Program Toolkit*. https://www.nsdassoc.ca/images/media/documents/2018-CCP-Toolkit.pdf (Accessed 15 January 2023).

Riediger, N. D., Kingson, O., Mudryj, A., Farquhar, K. L., Spence, K. A., Vagianos, K., & Suh, M. (2018). Diversity and equity in dietetics and undergraduate nutrition education in Manitoba. *Canadian Journal of Dietetic Practice and Research*, *80*(1), 44–46.

Rogers, D. (2000). Report on the American Dietetic Association's member needs assessment/satisfaction study. *Journal of the American Dietetic Association*, *1*(100), 112–116.

Rogers, D. (2005). Report on the American Dietetic Association/ADA foundation/Commission on Dietetic Registration 2004 dietetics professionals needs assessment. *Journal of the American Dietetic Association*, *105*(9), 1348–1355.

Rogers, D. (2009). Report on the American Dietetic Association/Commission on Dietetic Registration 2008 needs assessment. *Journal of the American Dietetic Association*, *7*(109), 1283–1293.

Rogers, D. (2013). Report on the Academy's 2012 needs satisfaction survey. *Journal of the Academy of Nutrition and Dietetics*, *113*(1), 146–152.

Rogers, D. (2017). Report on the Academy/Commission on Dietetic Registration 2016 needs satisfaction survey. *Journal of the Academy of Nutrition and Dietetics*, *4*(117), 626–631.

Rogers, D. (2021). From the Academy Report on the Academy/Commission on Dietetic Registration 2020 needs satisfaction survey. *Journal of the Academy of Nutrition and Dietetics, 121*(1), 134–138.

Schier, H. E., Gunther, C., Landry, M. J., et al. (2022). Sex and gender data collection in nutrition research: Considerations through an inclusion, diversity, equity and access lens. *Journal of the Academy of Nutrition and Dietetics*, S2212-2672.

Scott, K. M. (2009). *Recipe for citizenship: Professionalization and power in World War I dietetics.* The College of William and Mary.

Smedley, B. D., & Mittman, I. S. (2011). The diversity benefit: How does diversity among health professionals address public needs? In *Healthcare disparities at the crossroads with healthcare reform* (pp. 167–193). Boston, MA: Springer.

Statistics Canada. (2022, October 26). *The daily—The Canadian census: A rich portrait of the country's religious and ethnocultural diversity [Government].* Statistics Canada. https://www150. statcan.gc.ca/n1/daily-quotidien/221026/dq221026b-eng.htm.

Stein, K. (2012). The educational pipeline and diversity in dietetics. *Journal of the Academy of Nutrition and Dietetics, 112*(6), 791–800.

The Academy of Nutrition and Dietetics. https://www.dietitians.ca/Become-a-Dietitian/Education-and-Training.

Tufuga, D., Mueller, K., Bellini, S. G., Stokes, N., & Patten, E. V. (2022). Content analysis of websites of didactic programs in dietetics for evidence of diversity, equity, and inclusion. *Journal of Nutrition Education and Behavior, 54*(12), 1116–1124.

United States Census Bureau. *Quick Facts.* https://www.census.gov/quickfacts/fact/table/US/ IPE120221 (Accessed 12 January 2023).

White, J. H., & Beto, J. A. (2013). Strategies for addressing the internship shortage and lack of ethnic diversity in dietetics. *Journal of the Academy of Nutrition and Dietetics, 113*(6), 771–775.

Wilson, S. (2008). Research is ceremony. In *Indigenous research methods.* Winnipeg: Fernwood.

Chapter 4

Equity, diversity, and inclusion in medicine: Sisyphean undertaking or achievable reality

Muhammad Mustafa Memon[a], Jeffrey Ding[b], Marissa Joseph[c,d,e], Philip R. Doiron[f], and Faisal Khosa[b]

[a]*Department of Medicine, Rochester General Hospital, Rochester, NY, United States,* [b]*University of British Columbia, Vancouver, BC, Canada,* [c]*Department of Pediatrics, Faculty of Medicine, University of Toronto, ON, Canada,* [d]*Section of Dermatology, Division of Paediatric Medicine, The Hospital for Sick Children, Toronto, ON, Canada,* [e]*Division of Dermatology, Women's College Hospital, Toronto, ON, Canada,* [f]*Division of Dermatology, Department of Medicine, University of Toronto, Toronto, ON, Canada*

> *Of all the forms of inequality, injustice in health care is the most shocking and inhumane*
>
> —Martin Luther King Jr.

Health and access to healthcare are fundamental human rights, which have been plagued by inequity. Lack of equity, diversity, and inclusion (EDI) has been a long-standing concern within the healthcare professions. As the demographics of the global population continue to evolve and diversify, the healthcare workforce must reflect the demographics of the communities they serve. Historically, medicine has not been inclusive and welcoming to racialized groups, including students, practitioners, and patients. Lack of diversity within medicine not only affects the quality of care that patients receive but also perpetuates systemic inequalities and health disparities within marginalized communities (Fig. 1). In this chapter, we summarize the current state of workforce diversity in medicine, discuss historical developments, present a rationale for systemic action, and discuss evidence-based solutions and interventions for increasing representation.

Equity, Diversity, and Inclusion in Healthcare. https://doi.org/10.1016/B978-0-443-13251-3.00004-1

FIG. 1 This word cloud illustrates the multifaceted nature of health disparities, highlighting the numerous interconnected areas that contribute to inequities in health outcomes.

Current state of workforce diversity in medicine

Current trends in the physician workforce show a stark lack of diversity, particularly among certain demographic groups. According to American Medical Colleges (AAMC) data from 2019, only 5% of active physicians in the United States (US) were African American, and just under 6% were Hispanic or Latino (Figure 18. Percentage of all active physicians by race/ethnicity, 2018). These numbers are disproportionately low compared to the general population, in which African Americans and Hispanics/Latinos make up around 13.6% and 18.9%, respectively (U.S. Census Bureau, 2020). These underrepresented minorities (URM) along with American Indian/Alaska Native/Native Hawaiian/Pacific Islander (AI/AN/NH/PI) constitute approximately a third (34.1%) of the US population. However, URM are not adequately represented in the physician workforce, instead making up only 11% of active physicians in the US. Additionally, women form only 35.8% of the physician workforce in the US, despite accounting for nearly half (50.5%) of the US population (U.S. Census Bureau, 2020).

Similarly, women constitute 43% of the physician workforce in Canada, despite comprising 50.3% of the general population (Physician data centre, 2023; Population, female (% of total population)—Canada, 2022). It is notable that Canadian healthcare authorities do not collect data on the race and ethnicity of the healthcare providers, and therefore, racial and ethnic disparities cannot be assessed (Rizvic, 2020). Thus, although likely to be present, the true impact of racial and ethnic disparity cannot be quantified and, therefore, addressed.

In contrast, the United Kingdom (UK) seems to have a much more diverse physician workforce. According to data from the National Health Service (NHS), physicians from Black and mixed ethnic groups constitute 4.6% and 3.2% of the workforce, respectively (NHS workforce, 2020). These proportions surpass even those of the general public of 4.0% (Black) and 2.9% (Mixed ethnic groups) (Garlick, 2021). Similarly, the proportion of women physicians in the UK is comparable to that of the general population at 47.3% versus 51.0% (Roskams, 2022; The state of medical education and practice in the UK: The workforce report, 2019). This disparity, although present, is one of the lowest in the Western world.

Moreover, enhancing diversity within medical research has the potential to address the historical underrepresentation of diverse groups in clinical trials. This underrepresentation is frequently rooted in mistrust and reticence among minority subjects to participate in research. However, studies have found that minority individuals are just as likely as nonminority individuals to agree to participate in research, which suggests that the methods used to recruit participants may need improvement (In plain sight: Addressing indigenous-specific racism and discrimination in B.C. Health Care, 2020; Wendler et al., 2005). This lack of representation in clinical trials can exacerbate health disparities among these disadvantaged populations. By increasing diversity among medical researchers, efforts and resources are more likely to be directed toward increasing participation and representation of underrepresented communities in clinical trials.

Furthermore, having a diverse healthcare workforce can also improve the cultural competency of the medical profession. A diverse team of healthcare providers is better equipped to understand and address the unique cultural and social factors that can impact a patient's health. For example, a team that includes providers from different racial and ethnic backgrounds may be better able to understand and address the barriers to healthcare faced by immigrant or refugee communities. One report that highlights the inequities these groups face is the "In Plain Sight" report, which showed that Indigenous people of British Columbia were far more likely to face stereotyping, racism, discrimination, lack of access to healthcare, and ultimately poor outcomes (Hafeez et al., 2017). By ensuring a workforce that is inclusive of minority groups, such as the Indigenous people, these inequities can be effectively addressed, leading to better care and outcomes for these patients.

In addition to racial and ethnic diversity, there is also a need for diversity in terms of gender and sexual orientation in medicine. For instance, research has shown that lesbian, gay, bisexual, and transgender (LGBT) patients often face discrimination and stigma when seeking healthcare, which can lead to poor health outcomes (Gonzaga et al., 2020). Healthcare providers who possess knowledge and sensitivity toward the healthcare needs of LGBT individuals can facilitate access to care and improve health outcomes in this population. It is important to note that the dearth of appropriate questions regarding gender and sexual identity in most national and state-level surveys creates a significant obstacle in accurately assessing the number of LGBT individuals and their healthcare needs.

Possible solutions and interventions

A problem as multifaceted as this requires an equally multidimensional solution. While no one solution can tackle this herculean task, multiple frameworks and best practice guidelines have been published proposing evidence-based interventions to make a concerted effort to effectively address this issue

(Assessing Institutional Culture and Climate, 2013; South-Paul et al., 2013). Herein, we discuss several approaches to dealing with these issues, keeping in mind these recommendations.

Setting diversity as a priority

For diversity initiatives to succeed, it is imperative to embed diversity as a priority in medical schools, residency programs, and institutional leadership (Auseon et al., 2013; Milem, 2003). Besides emphasizing the improvement of diversity in institutional composition, the incorporation of diversity must pervade the institution's mission and vision, strategic plans, policies, curriculum, and extracurricular endeavors (Castillo-Page et al., 2012).

The AAMC recommends, as part of its Diversity 3.0 Framework, that institutional representatives engage in reflective questioning regarding areas of strength and opportunities for development (South-Paul et al., 2013). This can be undertaken by a single individual, such as the chief diversity and/or inclusion officer, or by a task force created specifically to identify key issues and formulate an action plan. Questions should revolve around diversity and inclusion and how it relates to institutional and social context, structures and policies, and human capital (South-Paul et al., 2013).

The culture of an institution also holds significant importance and is shaped by past experiences and the beliefs and actions the institution has implemented to support diversity and inclusion. The AAMC recommends that an in-depth analysis of institutional culture be carried out by a group comprising at least two individuals (South-Paul et al., 2013). Both quantitative (i.e., surveys) and qualitative (i.e., focus groups and interview measures) should be used to assess the culture of an institution.

In addition, it is crucial to examine existing information, such as policies, documents, and statistics (e.g., student enrollment data) to gain a deeper understanding of how the organization has approached and promoted diversity. Moreover, data should be gathered from a variety of sources within the organization, as well as from diverse groups. When conducting qualitative research, creating safe and inclusive spaces where individuals can openly share their experiences and perspectives without the fear of being judged, retribution, or censure is of the utmost importance. Lastly, it is important to remember that data related to diversity and inclusion should be kept confidential and anonymous (Outtz, 2002).

An effective approach to this includes analyzing departmental and residency data on the race and ethnicity of faculty and trainees and comparing it to national averages. This not only helps to secure a commitment from department leaders but also ensures that the necessary resources are allocated to support URM recruitment efforts (Auseon et al., 2013). Furthermore, institutions can establish a uniform message of embracing EDI across website content, verbal interactions, and published materials (Rosenkranz et al., 2021). EDI

December 3rd—International day of persons with disabilities: Removing visible and invisible barriers. (2022, November 16). Government of Canada. https://www.canada.ca/en/department-national-defence/maple-leaf/defence/2022/11/december-3-international-day.html.

Deville, C., Chapman, C. H., Burgos, R., Hwang, W.-T., Both, S., & Thomas, C. R. (2014). Diversity by race, Hispanic ethnicity, and sex of the United States medical oncology physician workforce over the past quarter century. *Journal of Oncology Practice, 10*(5), e328–e334. https://doi.org/10.1200/JOP.2014.001464.

Ding, J., Haq, A. F., Joseph, M., & Khosa, F. (2021). Disparities in pediatric clinical trials for acne vulgaris: A cross-sectional study. *Journal of the American Academy of Dermatology,* S0190962221026542. https://doi.org/10.1016/j.jaad.2021.10.013.

Ding, J., Joseph, M., Chawla, S., Yau, N., Khosa, Z., Khawaja, F., & Khosa, F. (2022). Disparities in psoriasis clinical trials: A cross-sectional analysis. *Journal of the American Academy of Dermatology, 87*(6), 1386–1389. https://doi.org/10.1016/j.jaad.2022.09.006.

Ding, J., Joseph, M., Yau, N., & Khosa, F. (2021). Underreporting of race and ethnicity in paediatric atopic dermatitis clinical trials: A cross-sectional analysis of demographic reporting and representation. *British Journal of Dermatology,* bjd.20740. https://doi.org/10.1111/bjd.20740.

Ding, J., Zhou, Y., Khan, M. S., Sy, R. N., & Khosa, F. (2021). Representation of sex, race, and ethnicity in pivotal clinical trials for dermatological drugs. *International Journal of Women's Dermatology, 7*(4), 428–434. https://doi.org/10.1016/j.ijwd.2021.02.007.

Doctors Back to School™ Day. (2023, October 20). American Medical Association. https://www.ama-assn.org/member-groups-sections/minority-affairs/doctors-back-school-day.

Doria, A., et al. (2024). Increasing diversity in Canadian radiology: From the hiring process to needed active retention efforts. *Canadian Assoication of Radiologists Journal.* https://pubmed.ncbi.nlm.nih.gov/38752404/.

Dossett, L. A., Mulholland, M. W., & Newman, E. A. (2019). Building high-performing teams in academic surgery: The opportunities and challenges of inclusive recruitment strategies. *Academic Medicine, 94*(8), 1142–1145. https://doi.org/10.1097/ACM.0000000000002647.

Feinberg, W. (2009). In H. LaFollette (Ed.), *Affirmative action* Oxford University Press. https://doi.org/10.1093/oxfordhb/9780199284238.003.0012.

Figure 13. Percentage of U.S. Medical School graduates by race/ethnicity (alone). (2019). AAMC. https://www.aamc.org/data-reports/workforce/interactive-data/figure-13-percentage-us-medical-school-graduates-race/ethnicity-alone-academic-year-2018-2019.

Figure 18. Percentage of all active physicians by race/ethnicity. (2018). AAMC. https://www.aamc.org/data-reports/workforce/data/figure-18-percentage-all-active-physicians-race/ethnicity-2018.

Garces, L. M., & Mickey-Pabello, D. (2015). Racial diversity in the medical profession: The impact of affirmative action bans on underrepresented student of color matriculation in medical schools. *The Journal of Higher Education, 86*(2), 264–294. https://doi.org/10.1080/00221546.2015.11777364.

Gardner, A. K. (2018). How can best practices in recruitment and selection improve diversity in surgery? *Annals of Surgery, 267*(1), e1–e2. https://doi.org/10.1097/SLA.0000000000002496.

Garlick, S. (2021). *Ethnic group, England and Wales: Census.* Office for National Statistics. https://www.ons.gov.uk/peoplepopulationandcommunity/culturalidentity/ethnicity/bulletins/ethnicgroupenglandandwales/census2021.

Gonzaga, A. M. R., Appiah-Pippim, J., Onumah, C. M., & Yialamas, M. A. (2020). A framework for inclusive graduate medical education recruitment strategies: Meeting the ACGME standard for a diverse and inclusive workforce. *Academic Medicine, 95*(5), 710–716. https://doi.org/10.1097/ACM.0000000000003073.

Gurin, P., Dey, E., Hurtado, S., & Gurin, G. (2002). Diversity and higher education: Theory and impact on educational outcomes. *Harvard Educational Review*, *72*(3), 330–367. https://doi.org/10.17763/haer.72.3.01151786u134n051.

Hafeez, D. M., Waqas, A., Majeed, S., Naveed, S., Afzal, K. I., Aftab, Z., Zeshan, M., & Khosa, F. (2019). Gender distribution in psychiatry journals' editorial boards worldwide. *Comprehensive Psychiatry*, *94*, 152119. https://doi.org/10.1016/j.comppsych.2019.152119.

Hafeez, H., Zeshan, M., Tahir, M. A., Jahan, N., & Naveed, S. (2017). Health care disparities among lesbian, gay, bisexual, and transgender youth: A literature review. *Cureus*. https://doi.org/10.7759/cureus.1184.

Hamidizadeh, R., Jalal, S., Pindiprolu, B., Tiwana, M. H., Macura, K. J., Qamar, S. R., Nicolaou, S., & Khosa, F. (2018). Influences for gender disparity in the radiology societies in North America. *AJR. American Journal of Roentgenology*, *211*(4), 831–838. https://doi.org/10.2214/AJR.18.19741.

Harley, E. H. (2006). The forgotten history of defunct black medical schools in the 19th and 20th centuries and the impact of the Flexner report. *Journal of the National Medical Association*, *98*(9), 1425–1429.

Hassouneh, D., Lutz, K. F., Beckett, A. K., Junkins, E. P., & Horton, L. L. (2014). The experiences of underrepresented minority faculty in schools of medicine. *Medical Education Online*, *19*(1), 24768. https://doi.org/10.3402/meo.v19.24768.

Heiser, S. (2019). *The majority of U.S. medical students are women, new data show*. AAMC. https://www.aamc.org/news/press-releases/majority-us-medical-students-are-women-new-data-show.

Hern, H. G., Alter, H. J., Wills, C. P., Snoey, E. R., & Simon, B. C. (2013). How prevalent are potentially illegal questions during residency interviews? *Academic Medicine*, *88*(8), 1116–1121. https://doi.org/10.1097/ACM.0b013e318299eecc.

Heron, S. L., Lovell, E. O., Wang, E., & Bowman, S. H. (2009). Promoting diversity in emergency medicine: Summary recommendations from the 2008 Council of Emergency Medicine Residency Directors (CORD) Academic Assembly Diversity Workgroup. *Academic Emergency Medicine*, *16*(5), 450–453. https://doi.org/10.1111/j.1553-2712.2009.00384.x.

Hutchinson, D., Das, P., Lall, M., Hill, J., Fares, S., & Khosa, F. (2021). Emergency medicine journal editorial boards: Analysis of gender, H-index, publications, academic rank, and leadership roles. *The Western Journal of Emergency Medicine*, *22*(2). https://doi.org/10.5811/westjem.2020.11.49122.

Important and commemorative days. (2017, October 16). Government of Canada. https://www.canada.ca/en/canadian-heritage/services/important-commemorative-days.html.

(2020). *In plain sight: Addressing indigenous-specific racism and discrimination in B.C. Health Care*. https://engage.gov.bc.ca/app/uploads/sites/613/2020/11/In-Plain-Sight-Summary-Report.pdf.

International women's day. (2024, March 18). Government of Canada. https://www.canada.ca/en/women-gender-equality/commemorations-celebrations/international-womens-day.html.

Jutras, M., Malekafzali, L., Jung, S., Das, P., Qamar, S. R., & Khosa, F. (2022). National Institutes of Health: Gender differences in radiology funding. *Academic Radiology*, *29*(5), 748–754. https://doi.org/10.1016/j.acra.2020.08.004.

Karol, D. L., Sheriff, L., Jalal, S., Ding, J., Larson, A. R., Trister, R., & Khosa, F. (2021). Gender disparity in dermatologic society leadership: A global perspective. *International Journal of Women's Dermatology*, *7*(4), 445–450. https://doi.org/10.1016/j.ijwd.2020.10.003.

Kim, K. Y., Kearsley, E. L., Yang, H. Y., Walsh, J. P., Jain, M., Hopkins, L., Wazzan, A. B., & Khosa, F. (2022). Sticky floor, broken ladder, and glass ceiling in academic obstetrics and gynecology in the United States and Canada. *Cureus*. https://doi.org/10.7759/cureus.22535.

Lu, J. D., Sverdlichenko, I., Siddiqi, J., & Khosa, F. (2021). Barriers to diversity and academic promotion in dermatology: Recommendations moving forward. *Dermatology, 237*(4), 489–492. https://doi.org/10.1159/000514537.

Ly, D. P., Essien, U. R., Olenski, A. R., & Jena, A. B. (2022). Affirmative action bans and enrollment of students from underrepresented racial and ethnic groups in U.S. Public Medical Schools. *Annals of Internal Medicine, 175*(6), 873–878. https://doi.org/10.7326/M21-4312.

Maddu, K., Amin, P., Jalal, S., Mauricio, C., Norbash, A., Ho, M.-L., Sanelli, P. C., Ali, I. T., Shah, S., Abujudeh, H., Nicolaou, S., Bencardino, J., & Khosa, F. (2021). Gender disparity in radiology society committees and leadership in North America and comparison with other continents. *Current Problems in Diagnostic Radiology, 50*(6), 835–841. https://doi.org/10.1067/j.cpradiol.2020.09.011.

Marder, A. (2017). *7 studies that prove the value of diversity in the workplace.* Capterra. https://www.capterra.com/resources/7-studies-that-prove-the-value-of-diversity-in-the-workplace/.

Mateo, C. M., & Williams, D. R. (2020). More than words: A vision to address bias and reduce discrimination in the health professions learning environment. *Academic Medicine, 95*(12S), S169–S177. https://doi.org/10.1097/ACM.0000000000003684.

McGaghie, W. C., Cohen, E. R., & Wayne, D. B. (2011). Are United States medical licensing exam step 1 and 2 scores valid measures for postgraduate medical residency selection decisions? *Academic Medicine, 86*(1), 48–52. https://doi.org/10.1097/ACM.0b013e3181ffacdb.

Michals, D. (2015). *Biography: Elizabeth Blackwell.* National Women's History Museum. https://www.womenshistory.org/education-resources/biographies/elizabeth-blackwell.

Milem, J. (2003). Chapter five The educational benefits of diversity: Evidence from multiple sectors. In M. J. Chang, D. Witt, J. Jones, & K. Hakuta (Eds.), *Compelling interest* (pp. 126–169). Stanford University Press. https://doi.org/10.1515/9780804764537-009.

Moghimi, S., Khurshid, K., Jalal, S., Qamar, S. R., Nicolaou, S., Fatima, K., & Khosa, F. (2019). Gender differences in leadership positions among academic nuclear medicine specialists in Canada and the United States. *American Journal of Roentgenology, 212*(1), 146–150. https://doi.org/10.2214/AJR.18.20062.

Mutwiri, G., Kulanthaivelu, R., Yuen, J., Hussain, M., Jutras, M., Deville, C., Jagsi, R., & Khosa, F. (2022). Gender differences among academic radiation oncology National Institutes of Health (NIH) funding recipients. *Cureus.* https://doi.org/10.7759/cureus.28982.

National Indigenous Peoples Day. (2024). Government of Canada. https://www.rcaanc-cirnac.gc.ca/eng/1100100013718/1708446948967.

Nfonoyim, B., Martin, A., Ellison, A., Wright, J. L., & Johnson, T. J. (2021). Experiences of underrepresented faculty in pediatric emergency medicine. *Academic Emergency Medicine, 28*(9), 982–992. https://doi.org/10.1111/acem.14191.

NHS workforce. (2020). Government of UK. https://www.ethnicity-facts-figures.service.gov.uk/workforce-and-business/workforce-diversity/nhs-workforce/latest/#main-facts-and-figures.

Outtz, J. (2002). The role of cognitive ability tests in employment selection. *Human Performance, 15*(1), 161–171. https://doi.org/10.1207/S15327043HUP1501&02_10.

Physician data centre. (2023). Canadian Medical Association.

Population, female (% of total population)—Canada. (2022). The World Bank. https://data.worldbank.org/indicator/SP.POP.TOTL.FE.ZS?locations=CA.

Preventza, O., Critsinelis, A., Simpson, K., Olive, J. K., LeMaire, S. A., Cornwell, L. D., Jimenez, E., Byrne, J., Chatterjee, S., Rosengart, T. K., & Coselli, J. S. (2021). Sex, racial, and ethnic disparities in U.S. cardiovascular trials in more than 230,000 patients. *The Annals of Thoracic Surgery, 112*(3), 726–735. https://doi.org/10.1016/j.athoracsur.2020.08.075.

Pride season—Canada.ca. (2024). Government of Canada. https://www.canada.ca/en/women-gender-equality/pride-season.html.

Queen's School of Medicine: Confronting exclusion. (2020). Queen's Alumni Review. https://www.queensu.ca/alumnireview/articles/2020-07-17/queen-s-school-of-medicine-confronting-exclusion.

Quick facts on Canada's physicians. (2019). Canadian Medical Association. https://www.cma.ca/quick-facts-canadas-physicians.

Riaz, I. B., Siddiqi, R., Zahid, U., Durani, U., Fatima, K., Sipra, Q.-A. R., Raina, A. I., Farooq, M. Z., Chamberlain, A. M., Wang, Z., Go, R. S., Marshall, A. L., & Khosa, F. (2020). Gender differences in faculty rank and leadership positions among hematologists and oncologists in the United States. *JCO Oncology Practice*, *16*(6), e507–e516. https://doi.org/10.1200/OP.19.00255.

Rizvic, S. (2020). *Why race-based data matters in health care—Institute for Canadian Citizenship.* Institute for Canadian Citizenship. https://inclusion.ca/article/why-race-based-data-matters-in-health-care/.

Rodríguez, J. E., Campbell, K. M., & Pololi, L. H. (2015). Addressing disparities in academic medicine: What of the minority tax? *BMC Medical Education*, *15*(1), 6. https://doi.org/10.1186/s12909-015-0290-9.

Rodríguez, J. E., Figueroa, E., Campbell, K. M., Washington, J. C., Amaechi, O., Anim, T., Allen, K.-C., Foster, K. E., Hightower, M., Parra, Y., Wusu, M. H., Smith, W. A., Villarreal, M. A., & Pololi, L. H. (2022). Towards a common lexicon for equity, diversity, and inclusion work in academic medicine. *BMC Medical Education*, *22*(1), 703. https://doi.org/10.1186/s12909-022-03736-6.

Rosenkranz, K. M., Arora, T. K., Termuhlen, P. M., Stain, S. C., Misra, S., Dent, D., & Nfonsam, V. (2021). Diversity, equity and inclusion in medicine: Why it matters and how do we achieve it? *Journal of Surgical Education*, *78*(4), 1058–1065. https://doi.org/10.1016/j.jsurg.2020.11.013.

Roskams, M. (2022). *Population and household estimates, England and Wales.* Office for National Statistics. https://www.ons.gov.uk/peoplepopulationandcommunity/populationandmigration/populationestimates/bulletins/populationandhouseholdestimatesenglandandwales/census2021unroundeddata.

Safdar, B., Naveed, S., Chaudhary, A. M. D., Saboor, S., Zeshan, M., & Khosa, F. (2021). Gender disparity in grants and awards at the National Institute of Health. *Cureus*. https://doi.org/10.7759/cureus.14644.

Shaikh, A. T., Farhan, S. A., Siddiqi, R., Fatima, K., Siddiqi, J., & Khosa, F. (2018). Disparity in leadership in neurosurgical societies: A global breakdown. *World Neurosurgery*. https://doi.org/10.1016/j.wneu.2018.11.145.

Sheikh, A. (2005). Why are ethnic minorities under-represented in US research studies? *PLoS Medicine*, *3*(2), e49. https://doi.org/10.1371/journal.pmed.0030049.

Sidhu, A., Jalal, S., & Khosa, F. (2020). Prevalence of gender disparity in professional societies of family medicine: A global perspective. *Cureus*. https://doi.org/10.7759/cureus.7917.

Simms, C. (2013). Voice: The importance of diversity in healthcare. *International Journal of Clinical Practice*, *67*(5), 394–396. https://doi.org/10.1111/ijcp.12134.

Smedley, B. D., Butler, A. S., Bristow, L. R., Institute of Medicine (U.S.), Committee on Institutional and Policy-Level Strategies for Increasing the Diversity of the U.S. Health Care Workforce, Institute of Medicine (U.S.), & Board on Health Sciences Policy. (2004). *In the nation's compelling interest: Ensuring diversity in the health-care workforce.* National Academies Press. http://site.ebrary.com/id/10062830.

South-Paul, J. E., Roth, L., Davis, P. K., Chen, T., Roman, A., Murrell, A., Pettigrew, C., Castleberry-Singleton, C., & Schuman, J. (2013). Building diversity in a complex academic health center. *Academic Medicine*, *88*(9), 1259–1264. https://doi.org/10.1097/ACM. 0b013e31829e57b0.

Starr, P. (2017). *The social transformation of American medicine (updated edition)*. Basic Books.

Steele, D. (2022). *Will more medical schools mean more black doctors?*. Inside Higher Ed. https://www.insidehighered.com/news/2022/05/13/new-med-schools-planned-need-black-doctors-continues.

Stephenson-Famy, A., Houmard, B. S., Oberoi, S., Manyak, A., Chiang, S., & Kim, S. (2015). Use of the interview in resident candidate selection: A review of the literature. *Journal of Graduate Medical Education*, *7*(4), 539–548. https://doi.org/10.4300/JGME-D-14-00236.1.

Teles, S. M. (1998). Why is there no affirmative action in Britain? *American Behavioral Scientist*, *41*(7), 1004–1026. https://doi.org/10.1177/0002764298041007010.

The state of medical education and practice in the UK: The workforce report. (2019). General Medical Council. https://www.gmc-uk.org/-/media/documents/the-state-of-medical-education-and-practice-in-the-uk- - -workforce-report_pdf-80449007.pdf.

Tien, C.-W., Tao, B., & Khosa, F. (2023). Gender disparity among ophthalmologists awarded Canadian institute of health research grants. *Women & Health*, *63*(2), 143–149. https://doi.org/10.1080/03630242.2022.2164113.

Tiffin, P. A., Dowell, J. S., & McLachlan, J. C. (2012). Widening access to UK medical education for under-represented socioeconomic groups: Modelling the impact of the UKCAT in the 2009 cohort. *BMJ*, *344*(apr17 2), e1805. https://doi.org/10.1136/bmj.e1805.

Toor, A. S., Wooding, D. J., Masud, S., & Khosa, F. (2021). Gender distribution in awarded Canadian Institutes of Health Research grants among anesthesiologists: A retrospective analysis between 2008 and 2020. *Canadian Journal of Anesthesia/Journal Canadien d'Anesthésie*, *68* (10), 1580–1581. https://doi.org/10.1007/s12630-021-02043-w.

Truesdale, C. M., Baugh, R. F., Brenner, M. J., Loyo, M., Megwalu, U. C., Moore, C. E., Paddock, E. A., Prince, M. E., Strange, M., Sylvester, M. J., Thompson, D. M., Valdez, T. A., Xie, Y., Bradford, C. R., & Taylor, D. J. (2021). Prioritizing diversity in otolaryngology–head and neck surgery: Starting a conversation. *Otolaryngology–Head and Neck Surgery*, *164*(2), 229–233. https://doi.org/10.1177/0194599820960722.

U.S. Census Bureau. (2020). *QuickFacts: United States*. https://www.census.gov/quickfacts/fact/table/US/POP010220.

Vassie, C., Smith, S., & Leedham-Green, K. (2020). Factors impacting on retention, success and equitable participation in clinical academic careers: A scoping review and meta-thematic synthesis. *BMJ Open*, *10*(3), e033480. https://doi.org/10.1136/bmjopen-2019-033480.

Vaughn, B. (2013, April 30). The history of diversity training & its pioneers. *Diversity Officer Magazine*. https://diversityofficermagazine.com/diversity-inclusion/the-history-of-diversity-training-its-pioneers/.

Walling, A., Nilsen, K., & Templeton, K. J. (2020). The only woman in the room: Oral histories of senior women physicians in a Midwestern City. *Women's Health Reports*, *1*(1), 279–286. https://doi.org/10.1089/whr.2020.0041.

Waseem, Y., Mahmood, S., Siddiqi, R., Usman, M. S., Fatima, K., Acob, C., & Khosa, F. (2019). Gender differences amongst board members of endocrinology and diabetes societies. *Endocrine*, *64*(3), 496–499. https://doi.org/10.1007/s12020-019-01861-9.

Wei, W., Cai, Z., Ding, J., Fares, S., Patel, A., & Khosa, F. (2023). Organizational leadership gender differences in medical schools and affiliated universities. *Journal of Women's Health*, jwh.2023.0326. https://doi.org/10.1089/jwh.2023.0326.

Wendler, D., Kington, R., Madans, J., Wye, G. V., Christ-Schmidt, H., Pratt, L. A., Brawley, O. W., Gross, C. P., & Emanuel, E. (2005). Are racial and ethnic minorities less willing to participate in health research? *PLoS Medicine*, *3*(2), e19. https://doi.org/10.1371/journal.pmed.0030019.

Whetzel, D. L., McDaniel, M. A., & Nguyen, N. T. (2008). Subgroup differences in situational judgment test performance: A meta-analysis. *Human Performance*, *21*(3), 291–309. https://doi.org/10.1080/08959280802137820.

Whitla, D. K., Orfield, G., Silen, W., Teperow, C., Howard, C., & Reede, J. (2003). Educational benefits of diversity in medical school: A survey of students. *Academic Medicine*, *78*(5), 460–466. https://doi.org/10.1097/00001888-200305000-00007.

Wu, B., Bhulani, N., Jalal, S., Ding, J., & Khosa, F. (2019). Gender disparity in leadership positions of general surgical societies in North America, Europe, and Oceania. *Cureus*, *11*(12), e6285. https://doi.org/10.7759/cureus.6285.

Wynn, R. (2000). Saints and sinners: Women and the practice of medicine throughout the ages. *JAMA*, *283*(5), 668. https://doi.org/10.1001/jama.283.5.668-JMS0202-4-1.

Young, P. J., Kagetsu, N. J., Tomblinson, C. M., Snyder, E. J., Church, A. L., Mercado, C. L., Guzman Perez-Carrillo, G. J., Jha, P., Guerrero-Calderon, J. D., Jaswal, S., Khosa, F., & Deitte, L. A. (2023). The intersection of diversity and well-being. *Academic Radiology*, *30*(9), 2031–2036. https://doi.org/10.1016/j.acra.2023.01.028.

examine race relations and how historical race relations continue to shape relations within the Canadian diverse society (Canadian Center for Diversity and Inclusion, 2022). We refer to diversity as the variety of unique dimensions, qualities, and characteristics that individuals possess (Canadian Center for Diversity and Inclusion, 2022). Such dimensions include socioeconomic status (including education, economic status, and other social factors), language, culture, ethnicity, gender, disability, ability, and age (Jefferies et al., 2019). Such dimensions can be visible or invisible to others.

While we define what diversity is and what it refers to, we also emphasize the importance of how individuals, organizations, and systems respond to diversity. While it is important to understand what diversity means, as well as the complex interaction of individuals' diverse characteristics, it is equally important to understand what to do with and how organizations respond to diversity. It is important to acknowledge that diversity also refers to how an organization or system affirms the uniqueness of and differences among individuals, families, community, and population (National League for Nursing, 2017). How nursing responds to the diversity of the nursing workforce is relevant to the discussion of EDI. Another important note related to the diversity of peoples is understanding how the different characteristics of individuals interact with one another and within the broader social context. This is often understood as the concept of *intersectionality* (Cho et al., 2013)—the complex interaction of individual characteristics and identities within an equally complex social landscape and how this interaction influences health and social outcomes of individuals.

In the latter sections of this chapter, we refer to diversity-related data. By this, we refer to data that provides information on the diverse characteristics of individuals and groups of individuals. This can include age, gender, sex, level of formal education, employment status, income, as well as race and/or ethnicity data. Although *race-based* and *ethnicity-based* data are often used interchangeably, we acknowledge the similarity of such concepts, as well as the important distinction between race and ethnicity. *Race* classifies groups of individuals based on perceived physical differences (e.g., skin color and facial features) tied to collective geographic or historical factors (Canadian Institute for Health Information (CIHI), 2020; Edmonton Social Planning Committee, 2021). *Ethnicity*, on the other hand, refers to a sense of group belonging based on shared factors, e.g., geographic spaces, cultural practices, language, or spirituality (Edmonton Social Planning Committee, 2021; Government of Canada, 2019). While very similar, these two terms remain very distinct from one another. The CIHI (2020) described that race-based data is helpful in responding to racial and biased inequalities within systems, while ethnicity data is helpful in program planning and delivery. Together, addressing both constructs is valuable in understanding systemic racism and discrimination, and collecting data on race and ethnicity provides important information to help address inequities within society (Edmonton Social Planning Committee, 2021).

Inclusion

Inclusion in society is a human right (United Nations. Human Rights Office of the High Commissioner, n.d.) and involves purposeful actions to promote participation, engagement, and a sense of belonging for members within society, particularly for individuals who have been historically underrepresented and disadvantaged. With respect to the nursing workforce, we examine issues of inclusion that include representation of diverse nurses in the workforce, particularly those who have been historically underrepresented or oppressed due to socioeconomic factors such as gender, colonialism, racism, and others. Indicators for inclusion include participation in the workforce, representation in decision-making positions, engagement and participation, health and wellbeing, income, social connections, education and development opportunities, and discrimination (Statistics Canada, 2023). Inclusion in the nursing workforce is critical in promoting the health and wellbeing of nurses, as well as the patients and clients that nurses serve through nursing practice.

Equity, diversity, inclusion in the nursing workforce: Taking stock

Historical context: Shaping the present state

We felt strongly about including a brief contextual piece on the history of nursing that continues to shape current issues related to EDI in the nursing workforce. As a female-dominated profession and discipline, a discussion on EDI in the nursing workforce cannot be complete without providing a context on the sex and gender differences that are present in nursing. In addition, as a profession that some argue was born out of the discipline and practice of medicine (MacMillan, 2012), power dynamics within the health sector continue to persist between nursing, medicine, and other health disciplines. It is our intention to provide an important historical context that has shaped and continues to shape issues of EDI in the nursing workforce today.

Nursing as a profession and discipline has a complicated history from both a global and national perspective. To add to the complexity of this history is the way it has been told, by whom, and in what time period. For example, the history of nursing is often a historical account of women's work as nurses despite the history of men working as nurses since the 4th century (Evans, 2004; Mackintosh, 1997).[a] We summarize selected historical factors that influence current EDI challenges in the nursing workforce. It is well beyond the scope of this chapter to present a fulsome discussion and synthesis of the complex

---

a. For this chapter, we will refer to the dichotomy of men/women or males/females in our gender-related discussions for several reasons. First, the literature refers to *men* or *women*, and less frequently, *males* or *females* in nursing. Second, our discussions represent what is reflected in the literature. Importantly, we recognize the diversity of gender identities which are fluid and dynamic, moving beyond the dichotomy of men/women.

enjoying full participation [in a group]" (Newman, 2014, p. 2). Globally, men continue to be underrepresented in the nursing workforce (WHO, 2019). However, researchers have demonstrated gender-related inequities within such disparities in the nursing workforce.

Interestingly, the gender imbalance in the nursing workforce also carries additional considerations in EDI in the nursing workforce. First, men, despite being underrepresented in nursing, excel more quickly than their female counterparts in the profession of nursing (Gauci et al., 2023). Gauci et al. (2023) found that men held a "privileged position within nursing, largely as a result of gender expectations" (p. 13). In another study, Gauci et al. (2022) found that men in nursing were perceived to have fewer consequences in the nursing workplace, be scrutinized less by other healthcare professionals, and be subject to fewer rules and regulations in the workplace. In another example, Smith et al. (2020) found that men in nursing felt more respected in the healthcare workplace by other healthcare providers than their female counterparts. Finally, research has shown that men in nursing tend to occupy senior nursing leadership roles in the nursing workforce and higher rates of pay when compared to their female nursing colleagues (Aspinall et al., 2021; Gauci et al., 2022; Mao et al., 2021; Smith et al., 2021).

Second, women in nursing often experience an interruption in their careers or lack of career advancement due to expectations of women to care for their families, including children and aging parents (Aspinall et al., 2021; Gauci et al., 2022; McIntosh, McQuaid, & Munro, 2015). Furthermore, nurses who returned to the nursing workforce after caring for family had less access to professional development opportunities or had to choose between a part-time and full-time nursing career to attend to caregiving of family dependents (McIntosh, Prowse, & Archibong, 2015).

Celebi Cakiroglu and Baykal (2021) asserted that perceptions and experiences of gender inequities in the nursing workforce are related to stress, burnout, decreased job satisfaction, and increased turnover in the workforce. Therefore, it is important to address not only the gender disparities within the nursing workforce, but also the inequities that emerge alongside such disparities.

Historically underrepresented individuals in nursing

In this section, we highlight the need to create a nursing workforce that is representative of the patient and client populations that nurses care for. We focus our discussion using Canada as an example, recognizing that the inclusion of underrepresented individuals in society in the nursing workforce will vary from country to country. In the Canadian context, underrepresented groups are referred to *equity-deserving groups*[b] including women, Black Peoples,

---

b. Equity-deserving groups is terminology that is often located in Canadian public policies which aim to address historic and systemic barriers and discrimination faced by members of these groups.

Indigenous Peoples, Persons of color (members of visible minority and racialized groups), Persons living with a disability, and members of LGBTQ2+ communities (Canadian Human Rights Commission, n.d.; Government of Canada, 2023).

Canada is known globally as a multicultural society; a country that is ethnoculturally and religiously diverse (Statistics Canada, 2022). In a population of 39.6 million people, there were more than 450 ethnic or cultural origins reported by Canadians. Over half of the Canadian population reported a Christian religion, while more than one-third of the population reported having no religious affiliation. In 2021, data demonstrated a growing population that reported being Muslim, Hindu, or Sikh. In terms of race-based data, Statistics Canada (2022) reported that one in four people in Canada is part of a racialized group. Three racialized groups represented 16.1% of the Canadian population: South Asian (7.1%), Chinese (4.7%), and Black Peoples (4.3%). Other racialized groups in Canada include Filipinos (2.6%), Arabs (1.9%), and Latin Americans (1.6%). Recent population projections for Canada demonstrate that the racialized population will continue to increase (Statistics Canada, 2022). Historically, the percentage of immigrants in the Canadian population has been nearly one-quarter of the total population (Statistics Canada, 2022). Furthermore, about 30% of the Canadian population is a visible minority. Despite the diversity of the Canadian population, the healthcare workforce in general and the nursing workforce in particular poorly reflects this diversity.

The nursing workforce remains a predominantly White and heteronormative female profession (Fontenot & McMurray, 2020; Jefferies et al., 2019; Merry et al., 2021). There is clear agreement among nursing scholars and organizations that diversity in the nursing profession will reduce disparities and inequities within the nursing workforce, but also promote excellence in patient and client care (Jefferies et al., 2019; Phillips & Malone, 2014; Premji & Etowa, 2014). For example, a diverse and inclusive nursing workforce contributes to accessible healthcare services, particularly for underrepresented populations in society, culturally appropriate healthcare services, improved patient-nurse relationships, and ultimately, improved health outcomes (Human Health Service Advisory Committee on Minority Health, 2011; Ong-Flaherty, 2015; Taylor et al., 2022).

Internationally educated nurses and nursing workforces

Cornelissen (2021) examined the profile of immigrants in Canadian nursing and health care support occupations, by intended occupation upon migration to and settlement in Canada, and by immigration category, as well as the professional integration of immigrants who completed their nursing education in Canada and overseas. Findings showed that immigrants who arrived in Canada as adults are overrepresented in nursing and health care support occupations, making up 28%

Therapy Advocates for Diversity (COTAD) are amplifying the issues and taking action. Essential to this ongoing evolution is the necessary recognition of occupational therapy as not simply a profession, but a global political *project* (Pollard et al., 2008). If structural and systemic change can be sustained, then occupational therapy might one day realize its potential as a holistic, just, and accountable space where all identities belong.

To conclude, the following section lists resources to support the diversification of the occupational therapy workforce:

Supporting Diversity in Higher Education:
https://www.cotad.org/cotaded
https://nbotc.wildapricot.org/
http://www.notpd.org/
Supporting Diversity in the Workplace:
https://www.aota.org/practice/practice-essentials/dei
https://www.jci.org/articles/view/141675
http://www.notpd.org/
https://nbotc.wildapricot.org/
Supporting Diversity in Healthcare Settings:
https://www.cotad.org/cotad-practitioners
https://www.aota.org/practice/practice-essentials/dei/transgender-and-gender-diverse-resources
https://www.aota.org/-/media/corporate/files/aboutot/dei/guide-racial-discrimination.pdf
https://caot.ca/site/adv-news/advocacy/trc-indigenous?nav=sidebar&banner=5

Disclosures

The authors report no conflicts of interest.

References

Abreu, B. C., & Peloquin, S. M. (2004). Embracing diversity in our profession. *American Journal of Occupational Therapy, 58*(3), 353–359. https://doi.org/10.5014/ajot.58.3.353.

American Occupational Therapy Association. (1943, March). *Minutes of the sub-Committee on schools and curriculum of the Educational Committee.*

American Occupational Therapy Association. (1991). Essentials and guidelines for an accredited educational program for the occupational therapist. *American Journal of Occupational Therapy, 45,* 1077–1084.

American Occupational Therapy Association. (2007). AOTA's centennial vision and executive summary. *American Journal of Occupational Therapy, 61*(6), 613–614. https://doi.org/10.5014/ajot.61.6.613.

American Occupational Therapy Association. (2017). Vision 2025. *American Journal of Occupational Therapy, 71*(3), 7103420010p1. https://doi.org/10.5014/ajot.2017.713002.

Glazer, G., Tobias, B., & Mentzel, T. (2018). Increasing healthcare workforce diversity: Urban universities as catalysts for change. *Journal of Professional Nursing, 34*(4), 239–244. https://doi.org/10.1016/j.profnurs.2017.11.009.

Gomez, L. E., & Bernet, P. (2019). Diversity improves performance and outcomes. *Journal of the National Medical Association, 111*(4), 383–392. https://doi.org/10.1016/j.jnma.2019.01.006.

Grenier, M.-L. (2020). Cultural competency and the reproduction of White supremacy in occupational therapy education. *Health Education Journal, 79*(6), 633–644. https://doi.org/10.1177/0017896920902515.

Grenier, M.-L., Zafran, H., & Roy, L. (2020). Current landscape of teaching diversity in occupational therapy education: A scoping review. *American Journal of Occupational Therapy, 74*(6), 7406205100p1–7406205100p15. https://doi.org/10.5014/ajot.2020.044214.

Guajardo Córdoba, A., & Galheigo, S. M. (2015). Reflexiones críticas sobre derechos humanos: Contribuciones de la Terapia Ocupacional Latinoamericana [Critical reflections about the human rights: Contributions from Latin America occupational therapy]. *World Federation of Occupational Therapists Bulletin, 71*(2), 73–80. https://doi.org/10.1179/1447382815Z.00000000023.

Guajardo, A., Kronenberg, F., & Ramugondo, E. L. (2015). Southern occupational therapies: Emerging identities, epistemologies and practices. *South African Journal of Occupational Therapy, 45*(1), 3–10. https://doi.org/10.17159/2310-3833/2015/v45no1a2.

Gustafsson, L., McKinstry, C., Buchanan, A., Laver, K., Pepin, G., Aplin, T., Hyett, N., Isbel, S., Liddle, J., & Murray, C. (2022). Doing, being, becoming, and belonging—A diversity, equity, and inclusion commitment. *Australian Occupational Therapy Journal, 69*(4), 375–378. https://doi.org/10.1111/1440-1630.12831.

Hammell, K. R. W. (2009). Sacred texts: A sceptical exploration of the assumptions underpinning theories of occupation. *Canadian Journal of Occupational Therapy, 76*(1), 6–13.

Hammell, K. W. (2011). Resisting theoretical imperialism in the disciplines of occupational science and occupational therapy. *British Journal of Occupational Therapy, 74*(1), 27–33. https://doi.org/10.4276/030802211X12947686093602.

Hammell, K. W. (2019). Building globally relevant occupational therapy from the strength of our diversity. *World Federation of Occupational Therapists Bulletin, 75*(1), 13–26. https://doi.org/10.1080/14473828.2018.1529480.

Hammell, K. W. (2023). Focusing on "what matters": The occupation, capability and wellbeing framework for occupational therapy. *Brazilian Journal of Occupational Therapy, 31*, e3509. https://doi.org/10.1590/2526-8910.ctoAO269035092.

Iwama, M. K. (2006). *The Kawa model: Culturally relevant occupational therapy*. Churchill Livingstone Elsevier.

Jaegers, L. A., Muñoz, J., Washington, S., Rogers, S., Skinner, E., Dieleman, C., White, J. A., Haworth, C., Shea, C.-K., Conners, B., Millsap, M., Young, C., West-Bruce, S., Hennessy, A., Nelson, M., Dillon, M. B., Daaleman, C., & Barney, K. F. (2020). Justice-based occupational therapy response to occupational therapy statements on justice and racism. *Justice-Based Occupational Therapy, 2*(2). https://www.slu.edu/mission-identity/initiatives/transformative-justice/pdfsimages/jbot-newsletter-vol2-issue2.pdf.

Jesus, T. S., Mani, K., von Zweck, C., Bhattacharjya, S., Kamalakannan, S., & Ledgerd, R. (2023). The global status of occupational therapy workforce research worldwide: A scoping review. *American Journal of Occupational Therapy, 77*(3), 7703205080. https://doi.org/10.5014/ajot.2023.050089.

Jesus, T. S., Mani, K., von Zweck, C., Kamalakannan, S., Bhattacharjya, S., Ledgerd, R., & on behalf of the World Federation of Occupational Therapists. (2022). Type of findings generated by the occupational therapy workforce research worldwide: Scoping review and content analysis. *International Journal of Environmental Research and Public Health*, *19*(9), 5307. https://doi.org/10.3390/ijerph19095307.

Johnson, K. R., Kirby, A., Washington, S., Lavalley, R., & Faison, T. (2022). Linking antiracist action from the classroom to practice. *American Journal of Occupational Therapy*, *76*(5), 7605347010.

Karaba Bäckström, M., Luiz Moura De Castro, A., Eakman, A. M., Ikiugu, M. N., Gribble, N., Asaba, E., Kottorp, A., Falkmer, O., Eklund, M., Ness, N. E., Balogh, S., Hynes, P., & Falkmer, T. (2023). Occupational therapy gender imbalance; revisiting a lingering issue. *Scandinavian Journal of Occupational Therapy*, *30*(7), 1113–1121. https://doi.org/10.1080/11038128.2023.2220912.

Kendi, I. X. (2016). *Stamped from the beginning: The definitive history of racist ideas in America*. Nation Books.

Krishnakumaran, T., Bhatt, M., Kiriazis, K., & Giddings, C. E. (2022). Exploring the role of occupational therapy and forced migration in Canada. *Canadian Journal of Occupational Therapy*, *89*(3), 238–248. https://doi.org/10.1177/00084174221084463.

Kronenberg, F. (2013). Doing well-doing right together: A practical wisdom approach to making occupational therapy matter. *New Zealand Journal of Occupational Therapy*, *60*(1), 24–32.

Llorens, L. (1970). Facilitating growth and development: The promise of occupational therapy. The 1969 Eleanor Clark Slagle Lecture. *American Journal of Occupational Therapy*, *24*, 93–101.

Lopes, R. E., & Malfitano, A. P. S. (2021). *Social occupational therapy: Theoretical and practical designs*. Elsevier.

Ma, A., Sanchez, A., & Ma, M. (2019). The impact of patient-provider race/ethnicity concordance on provider visits: Updated evidence from the medical expenditure panel survey. *Journal of Racial and Ethnic Health Disparities*, *6*(5), 1011–1020. https://doi.org/10.1007/s40615-019-00602-y.

Mahoney, W. J., & Kiraly-Alvarez, A. F. (2019). Challenging the status quo: Infusing non-western ideas into occupational therapy education and practice. *Open Journal of Occupational Therapy*, *7*(3), 1–10. https://doi.org/10.15453/2168-6408.1592.

Malfitano, A. P. S. (2022). An anthropophagic proposition in occupational therapy knowledge: Driving our actions towards social life. *World Federation of Occupational Therapists Bulletin*, *78*(2), 70–82. https://doi.org/10.1080/14473828.2022.2135065.

Malfitano, A. P. S., da Mota de Souza, R. G., Townsend, E. A., & Lopes, R. E. (2019). Do occupational justice concepts inform occupational therapists' practice? A scoping review. *Canadian Journal of Occupational Therapy*, *86*(4), 299–312. https://doi.org/10.1177/0008417419833409.

Malfitano, A. P. S., Leite Junior, J. D., Bortolai, L. A., Farias, M. N., & Silva, M. J. (2022). Juventude e vulnerabilidade: Assessoria à assistência estudantil pela terapia ocupacional social [Youth and vulnerability: Assistance to student assistance through social occupational therapy]. In R. E. Lopes, & P. L. O. Borba (Eds.), *Terapia ocupacional, Educação e Juventudes: Conhecendo práticas e reconhecendo saberes [Occupational therapy, education and youth: Knowing practices and recognizing knowledge]* (pp. 433–448). São Carlos, Brazil: EdUFSCar.

McRae, E. G. (2018). *Mothers of massive resistance: White women and the politics of white supremacy*. Oxford University Press. https://doi.org/10.1093/oso/9780190271718.001.0001.

Melo, K. M. M., Malfitano, A. P. S., & Lopes, R. E. (2020). The social markers of the difference: Contributions to social occupational therapy. *Brazilian Journal of Occupational Therapy*, *28*(3), 1061–1071. https://doi.org/10.4322/2526-8910.ctoARF1877.

Meyer, A. (1922). The philosophy of occupational therapy. *Archives of Occupational Therapy, 1*, 1–10.

Monzeli, G. A., Morrison, R., & Lopes, R. E. (2019). Histories of occupational therapy in Latin America: The first decade of creation of the education programs. *Brazilian Journal of Occupational Therapy, 27*(2), 235–250. https://doi.org/10.4322/2526-8910.ctoAO1631.

Morán, J. P., & Ulloa, F. (2016). Perspectiva crítica desde latinoamérica: hacia una desobediencia epistémica en terapia ocupacional contemporánea [Critical perspective from Latin America: An epistemic disobedience in the contemporary Occupational Therapy]. *Brazilian Journal of Occupational Therapy, 24*(2), 421–427. https://doi.org/10.4322/0104-4931.ctoARF0726.

Muñoz, J. P. (2007). Culturally responsive caring in occupational therapy. *Occupational Therapy International, 14*(4), 256–280. https://doi.org/10.1002/oti.238.

Occupational Therapy Australia. (2016). OT AUSTRALIA position statement: Occupational deprivation. *Australian Occupational Therapy Journal, 63*(6), 445–447. https://doi.org/10.1111/1440-1630.12347.

Otte, S. V. (2022). Improved patient experience and outcomes: Is patient–provider concordance the key? *Journal of Patient Experience, 9*, 237437352211030. https://doi.org/10.1177/23743735221103033.

Padilla, R., & Griffiths, Y. (2017). *A professional legacy: The Eleanor Clarke Slagle lectures in occupational therapy, 1955-2016, Centennial Edition*. Retrieved July 13, 2023, from https://myaota.aota.org/shop_aota/product/900453U.

Palacios Tolvett, M. (2017). Reflexiones sobre las prácticas comunitarias: Aproximación a una Terapia Ocupacional del Sur [Reflections about the comunitario practices: Aproares from a South Occupational Therapy]. *Revista Ocupación Humana [Human Occupation Journal], 17*(1), 73–88. https://doi.org/10.25214/25907816.157.

Pereira, R. B., Whiteford, G., Hyett, N., Weekes, G., Di Tommaso, A., & Naismith, J. (2020). Capabilities, Opportunities, Resources and Environments (CORE): Using the CORE approach for inclusive, occupation-centred practice. *Australian Occupational Therapy Journal, 67*(2), 162–171. https://doi.org/10.1111/1440-1630.12642.

Peters, C. O. (2011). Powerful occupational therapists: A community of professionals, 1950–1980. *Occupational Therapy in Mental Health, 27*(3–4), 199–410. https://doi.org/10.1080/0164212X.2011.597328.

Pittman, P., Chen, C., Erikson, C., Salsberg, E., Luo, Q., Vichare, A., Batra, S., & Burke, G. (2021). Health workforce for health equity. *Medical Care, 59*(Suppl 5), S405–S408. https://doi.org/10.1097/MLR.0000000000001609.

Pollard, N., Sakellariou, D., & Kronenberg, F. (Eds.). (2008). *A political practice of occupational therapy* Elsevier Health Sciences.

Pooley, E. A., & Beagan, B. L. (2021). The concept of oppression and occupational therapy: A critical interpretive synthesis. *Canadian Journal of Occupational Therapy, 88*(4), 407–417. https://doi.org/10.1177/00084174211051168.

Pooley, E., Lizon, M., & Sangrar, R. (2023). Towards equity and justice in occupational therapy. *Occupational Therapy Now, 26*(1), 20–22. https://caot.ca/document/7946/Pages%20from%20Occupational%20Therapy%20Now%20January%202023-3.pdf.

Porsha, C., & Stromquist, N. P. (2015). Academic and diversity consequences of affirmative action in Brazil. *Compare: A Journal of Comparative and International Education, 45*(5), 792–813. https://doi.org/10.1080/03057925.2014.907030.

Quiroga, V. A. (1995). *Occupational therapy history: The first 30 years, 1900 to 1930*. American Occupational Therapy Association Press.

Royal College of Occupational Therapists. (2020). *RCOT statement on diversity*. https://www.rcot.co.uk/news/rcot-statement-diversity.

Salsberg, E., Richwine, C., Westergaard, S., Portela Martinez, M., Oyeyemi, T., Vichare, A., & Chen, C. P. (2021). Estimation and comparison of current and future racial/ethnic representation in the US health care workforce. *JAMA Network Open, 4*(3), e213789. https://doi.org/10.1001/jamanetworkopen.2021.3789.

Salvant, S., Kleine, E. A., & Gibbs, V. D. (2021). Be heard—We're listening: Emerging issues and potential solutions from the voices of BIPOC occupational therapy students, practitioners, and educators. *American Journal of Occupational Therapy, 75*(6), 7506347010. https://doi.org/10.5014/ajot.2021.048306.

Scaffa, M. E., & Reitz, S. M. (2020). *Occupational therapy in community and population health practice* (3rd ed.). F.A. Davis.

Sharma, S., Patlas, M., Khosa, F., & Yong-Hing, C. J. (2023). Equity, diversity and inclusion in radiology: Prioritizing trainee involvement. *Canadian Association of Radiologists journal = Journal l'Association Canadienne des Radiologistes*, 8465371231170230. https://doi.org/10.1177/08465371231170230.

Shen, M. J., Peterson, E. B., Costas-Muñiz, R., Hernandez, M. H., Jewell, S. T., Matsoukas, K., & Bylund, C. L. (2018). The effects of race and racial concordance on patient-physician communication: A systematic review of the literature. *Journal of Racial and Ethnic Health Disparities, 5*(1), 117–140. https://doi.org/10.1007/s40615-017-0350-4.

Taff, S. D. (2020). Educational philosophies influencing the development of early occupational therapy curricula. In S. D. Taff, L. C. Grajo, & B. L. Hooper (Eds.), *Perspectives on occupational therapy education: Past, present, and future* (pp. 3–11). Slack, Incorporated.

Taff, S. D., & Babulal, G. M. (2021). Reframing the narrative on the philosophies influencing the development of occupational therapy. In S. D. Taff (Ed.), *Philosophy and occupational therapy: Informing education, research, and practice* (pp. 7–22). Slack Incorporated.

Taff, S. D., & Blash, D. (2017). Diversity and inclusion in occupational therapy: Where we are, where we must go. *Occupational Therapy in Health Care, 31*(1), 72–83. https://doi.org/10.1080/07380577.2016.1270479.

Taff, S. D., & Clifton, M. (2022). Inclusion and belonging in higher education: A scoping study of contexts, barriers, and facilitators. *Higher Education Studies, 12*(3), 122. https://doi.org/10.5539/hes.v12n3p122.

Taff, S. D., & Putnam, L. (2022). Northern philosophies and professional neocolonialism in occupational therapy: A historical review and critique. *Brazilian Journal of Occupational Therapy, 30*, e2986.

Takeshita, J., Wang, S., Loren, A. W., Mitra, N., Shults, J., Shin, D. B., & Sawinski, D. L. (2020). Association of racial/ethnic and gender concordance between patients and physicians with patient experience ratings. *JAMA Network Open, 3*(11), e2024583. https://doi.org/10.1001/jamanetworkopen.2020.24583.

Taylor, K. J., Ford, L., Allen, E. H., Mitchell, F., Eldridge, M., & Caraveo, C. A. (2022, May). *Improving and expanding programs to support a diverse health care workforce*. Urban Institute. https://www.urban.org/sites/default/files/2022-05/Improving%20and%20Expanding%20Programs%20to%20Support%20a%20Diverse%20Health%20Care%20Workforce%20.pdf.

Testa, D. (2012). Aportes para el debate sobre los inicios de la profesionalización de la terapia ocupacional en Argentina [Contributions to the debate about the beginnings of the professionalization of occupational therapy in Argentina]. *Revista Chilena de Terapia Ocupacional [Chilean Journal of Occupational Therapy], 12*(1), 72–87. https://doi.org/10.5354/0719-5346.2012.22054.

United Nations. (2015). *The 17 sustainable development goals.* https://sdgs.un.org/goals.

Valantine, H. A., & Collins, R. S. (2015). National Institutes of Health addresses the science of diversity. *The Proceedings of the National Academy of Sciences, 112*(40), 12240–12242. https://doi.org/10.1073/pnas.1515612112.

Whiteford, G., Jones, K., Rahal, C., & Suleman, A. (2018). The participatory occupational justice framework as a tool for change: Three contrasting case narratives. *Journal of Occupational Science, 25*(4), 497–508. https://doi.org/10.1080/14427591.2018.1504607.

Wilbur, K., Snyder, C., Essary, A. C., Reddy, S., Will, K. K., & Saxon, M. (2020). Developing workforce diversity in the health professions: A social justice perspective. *Health Professions Education, 6*(2), 222–229. https://doi.org/10.1016/j.hpe.2020.01.002.

Williams, J. S., Walker, R. J., & Egede, L. E. (2016). Achieving equity in an evolving healthcare system: Opportunities and challenges. *The American Journal of the Medical Sciences, 351*(1), 33–43. https://doi.org/10.1016/j.amjms.2015.10.012.

World Federation of Occupational Therapists. (2019). *Occupational therapy and human rights— Position statement.* Retrieved from https://wfot.org/resources/occupational-therapy-and-human-rights. (Accessed 7 August 2023).

Xierali, I. M., & Nivet, M. A. (2018). The racial and ethnic composition and distribution of primary care physicians. *Journal of Health Care for the Poor and Underserved, 29*(1), 556–570. https://doi.org/10.1353/hpu.2018.0036.

Yong-Hing, C. J., & Khosa, F. (2023). Provision of culturally competent healthcare to address healthcare disparities. *Canadian Association of Radiologists journal = Journal l'Association Canadienne des Radiologistes, 74*(3), 483–484. https://doi.org/10.1177/08465371231154231.

Yong-Hing, C. J., Vaqar, M., Sahi, Q., & Khosa, F. (2023). Burnout: Turning a crisis into an opportunity. *Canadian Association of Radiologists journal = Journal l'Association Canadienne des Radiologistes, 74*(1), 16–17. https://doi.org/10.1177/08465371221130683.

Young, P. J., Kagetsu, N. J., Tomblinson, C. M., Snyder, E. J., Church, A. L., Mercado, C. L., Guzman Perez-Carrillo, G. J., Jha, P., Guerrero-Calderon, J. D., Jaswal, S., Khosa, F., & Deitte, L. A. (2023). The intersection of diversity and well-being. *Academic Radiology, 30*(9), 2031–2036. https://doi.org/10.1016/j.acra.2023.01.028.

Chapter 7

Seeing clearly: Equity, diversity, and inclusion in optometry

Gary Y. Chu[a], Crystal Lewandowski[a,b], Lillian Kalaczinski[c], Simone Jadczak[a], and Addy Rose[a]
[a]*New England College of Optometry, Boston, MA, United States, [b]North End Waterfront Health, Boston, MA, United States, [c]Michigan College of Optometry at Ferris State University, Big Rapids, MI, United States*

> *There must exist a paradigm, a practical model for social change that includes an understanding of ways to transform consciousness that are linked to efforts to transform structures.*
>
> bell hooks

> *I am no longer accepting the things I cannot change. I am changing the things I cannot accept.*
>
> Angela Y. Davis

Introduction

Historically, groups underrepresented in medicine are also underrepresented in optometry. For the majority of the time the profession of optometry has existed, diversity has not been a priority for the profession, nor have great strides been made to ensure representation in all its dimensions of society. While most other health professions have written about the importance of equity, diversity, and inclusion (EDI), optometry has unfortunately lagged behind (Yashadhana et al., 2021).

There have always been EDI champions and, in 2004, led by a small group of visionaries in the profession through the Association of Schools and Colleges of Optometry (ASCO), "A Road Map for Diversity in Optometric Education and the Profession" was developed (*A road map for diversity in optometric education and the profession*, 2004). The road map called for schools and colleges of optometry to "create, foster, and maintain an institutional climate that welcomes and embraces diversity" by conducting multicultural symposia and developing cultural competence/cultural communication seminars for faculty and staff. This started a Diversity and Cultural Competency Task Force with

Equity, Diversity, and Inclusion in Healthcare. https://doi.org/10.1016/B978-0-443-13251-3.00007-7

a vision of achieving diversity and cultural competence in optometric education and patient care. It was the first concerted national effort to explore and address the issues of EDI, which led ASCO to formalize the task force as a permanent standing committee that continues to this date.

ASCO's Diversity and Cultural Competency Committee has educated and trained numerous students, faculty, and developed leaders who have taken prominent roles in this important area of work. Like many academic disciplines unfortunately, it took the death of George Floyd to ignite the value and importance of EDI for optometry. But without the work of this standing committee in its foresight to create symposia, curriculum, advocacy, and research, the profession would not have made the progress it has. This chapter will explore the gaps that exist and encourage the profession of Optometry to collectively make progress and continue this journey.

Definition of optometry and its scope

According to the World Council of Optometry (WCO), "Optometry is a healthcare profession that is autonomous, educated, and regulated (licensed/registered), and optometrists are the primary healthcare practitioners of the eye and visual system who provide comprehensive eye and vision care, which includes refraction and dispensing, detection/diagnosis and management of diseases of the eye, and the rehabilitation of conditions of the visual system." Since it is a regulated profession, each country's government dictates the scope of practice, educational requirements, and licensure requirements. In some countries, like the United States (US) and Canada, an optometrist receives a Doctor of Optometry degree; in most other countries, optometry is a baccalaureate degree. Additionally, in some countries, such as the United Kingdom (UK), optometrists play a major role in an individual's overall health by detecting systemic diseases, diagnosing, and managing ocular manifestations of those diseases, and communicating with other healthcare professionals to coordinate health care for individuals.

In 2005, the WCO adopted the Global Competency-Based Model of Scope of Practice in Optometry, consisting of four different levels of competency, each of which builds upon and requires knowledge of all previous levels. Level 1 is characterized by optical services, including management and dispensing of ophthalmic lenses, frames, and other optical devices. Level 2 includes "investigation, examination, measurement, diagnosis, and correction/management of defects of the visual system." Level 3 includes "investigation, examination, and evaluation of the eye and adnexa and associated systemic factors to detect, diagnose, and manage disease." Level 4 includes ocular therapeutic services, including "the use of pharmaceutical agents and other procedures to manage ocular conditions/disease." Of note, countries with level 4 competency at present include Australia, Canada, Columbia, New Zealand, the UK, and the US, and those with level 1 competency level include Turkey (Kuehn, 2021).

WHO competency level	Competency level definition	Example countries
1	Limited to optical services (management and dispensing of ophthalmic lenses, frames, and other optical devices)	Turkey
2	Level 1+examination, measurement, diagnosis, and correction/management of "defects of the visual system"	France, Greece, Hungary, Italy Poland, Slovenia
3	Level 1, 2+examination and evaluation of the "eye and adnexa and associated systemic factors to detect, diagnose, and manage disease"	Denmark, Germany, Spain, Portugal
4	Level 1, 2, 3+ocular therapeutic services including "the use of pharmaceutical agents and other procedures to manage ocular conditions/disease"	Australia, Canada, Columbia, New Zealand, the United Kingdom, and the United States of America

The global burden of eye disease makes it imperative that the number of optometrists and their ability to serve patients from diverse cultural backgrounds meet community needs. Since it is a relatively new profession, optometry is often misunderstood and relatively unknown to the majority of the population; hence, the workforce's diversity does not reflect the diversity of the demographics they serve.

Often, people associate optometry solely with vision correction, recalling personal anecdotes of getting eye glasses prescriptions or ordering contact lenses. This is only a fraction of the optometrists' scope of practice in providing eye care and preserving ocular health. Optometrists work in private practices, community health centers, academia, and industry. It is a multifaceted profession with a much broader scope of practice than many realize. It profoundly impacts individuals and communities alike, contributing to academic performance and occupational success. Many people owe their independence and quality of life to their optometrist.

Optometrists cater to patients of all ages, providing pediatric and geriatric care. The profession helps promote overall eye health by providing preventive care through a comprehensive eye examination. Their multifaceted role includes addressing refractive errors, as well as diagnosing, managing, and preventing eye conditions, such as glaucoma, cataracts, macular degeneration, and dry eye. Early prevention is one of the most essential components of the profession. Annual eye examinations allow for the identification of potential vision issues and monitoring changes in eye health over time. Unfortunately, many in the US and beyond do not receive the preventive screenings needed to detect and address emerging issues before they progress to more advanced stage (Kuehn, 2021).

Optometry is a regulated profession in the US, which mandates each state to govern the standards, regulations, and scope of practice for optometrists. The eye care community regularly engages with policymakers to advocate for an expanded scope of practice to address emerging challenges in eye care. Many optometrists advocate for laws allowing them to diagnose and treat eye conditions independently. For example, Massachusetts passed a bill in 2021 allowing optometrists (ODs) to use and prescribe topical and oral therapeutic agents for diagnosing, preventing, treating, or managing glaucoma (*Massachusetts scope win adds glaucoma authority*, 2021). These legislative victories ensure that patients receive comprehensive and timely care since Optometry is a patient-centered profession proactive in serving the visual health needs of individuals and communities.

Optometric care includes a variety of services to enhance eye health and vision. Optometrists bring a comprehensive approach to patient care, addressing not only refractive errors but also diagnosing, managing, and preventing eye conditions. Their expertise directly impacts individuals' well-being, enabling them to lead fulfilling lives with improved vision and ocular health.

Characterization of the present state of workforce diversity

Workforce demographics (age, race, gender)

In the US, 52.5% of optometrists are men, and 73.4% are White (non-Hispanic), making these the most common gender and race in the occupation. Minorities are poorly represented; 18.8% of optometrists report Asian race or ethnicity, and only 2% of optometrists identify Black race or ethnicity, compared to 5.9% and 13.4% of the US population, respectively. Additionally, only 3.8% of optometrists self-identified as Hispanic or Latino, compared to 18.7% of the population, and 0.2% identified as Native American, compared to 0.9% in the US (*Optometry's reflection*, 2021; Shneor et al., 2021). Over the next decade, the optometric workforce is projected to grow by about 1.4% each year (Cole, 2021).

In the UK, 62.6% of optometrists are women, 2% identified LBTQ+, and the remaining identified as men. In regard to ethnicity, 49.0% identified as White, 33% as Asian, 14% preferred not to disclose, and the remaining were black or mixed ethnicity groups (*General Optical Council: Equity and Diversity Data Monitoring Report*, 2021).

In Australia, 59.1% of all optometrists identified as women as of 2023, and the remaining identified as men. 0.2% of all optometrists identify as Aboriginal and/or Torres Strait Islander; however, there was no data about other races and ethnicities (*Optometry Board of Australia—Statistics*, 2023).

In Canada, as of 2016, 53% of Canadian optometrists were women with the province of Québec having the highest proportion of women optometrists, at 68% (Emily McMorris, 2016).

In Africa as of 2008, 57.8% of all registered optometrists were women, and 51.7% identify as White, 22.2% Black, and 21.9% Indian (Nirghin et al., 2011).

In New Zealand, 61.3% of optometrists were women. In New Zealand, the term "race" is typically avoided and only ethnicity is used (Harewood & Rosenfield, 2021). The ethnicities reported by most optometrists in New Zealand include European ethnicity (45.3%) and Asian (42.4%), followed by 3.3% reporting Middle Eastern/Latin America/Africa (MELAA) ethnicity and 2% Māori (*The optometrist and dispensing optician workforce in Aotearoa New Zealand*, 2023).

Student demographics

The low representation of optometrists of color predestined by low representation of optometry students and faculty of color—in part—contributes to inequities in eye care. The number of Black and Hispanic applicants using the Optometry Centralized Application Service (OptomCAS) increased by 14.3% and 19.4%, respectively, between the application cycles of 2019–20 and 2020–21. Thus, Black and Hispanic enrollment increased by 17.6% and 8.3% in 2020–21 and 4.0% and 7.8% in 2021–22. Since the 2017–18 academic year, the percentage of Native American and Alaska Native optometry students has increased from 0.5% to 0.6%. The percentage of Hawaiian or other Pacific Islander optometry students has remained unchanged at 0.2% (Elder, 2022).

In the UK, 65% of students identified as women in academic Year 2020, and this has not significantly changed over the past few years. The proportion of White students (36.0%) is lower than the optometry workforce profile. The percentage of Asian students is 50.4%, and other ethnicities remain underrepresented (*General Optical Council: Equity and Diversity Data Monitoring Report*, 2021).

Leadership demographics within the optometric profession

While progress has been made to diversify leadership roles in eye care, there is still a lack of data on the racial and ethnic diversity of optometry leadership in the US.

Though women are far less likely to hold positions of leadership in academic institutions, serving as directors, department chairs, or chief academic officers, optometric leadership is finally mirroring the change in the gender diversity reflected in the profession: 39% of deans of the schools and colleges of optometry in the US are women; 45% of the AOA Board of Trustees members are women; and 50% of the Board of Directors for the AAO are women (Elder, 2022).

The World Council of Optometry has geographic representatives from Africa, Europe, Asia Pacific, Latin America, and North America (*Board of Directors—World Council of Optometry*, 2024).

Diversity within optometric research, grants agencies, and journal editorial boards

According to an editorial in the Optometry and Vision Science Journal, to diversify their editorial board, they must develop specific plans on how to achieve program diversity (Twa, 2020). This includes initial recognition of a lack of EDI, even if unintentional, data collection and comparison to benchmarks of other similar organizations, and determination of specific, attainable goals to increase EDI—while engaging the community in conversations to promote transparency and accountability.

Though much of the information in this section focuses on race and gender, many other facets of diversity, including intersectionality between gender identity, religious beliefs, and socioeconomic status, should be considered when considering diversifying the optometric profession.

Longitudinal trends

United States

In the past 50 years, women in optometry have made huge strides in both their representation in the profession and in leadership roles within optometry. In 1899, Gertrude Stanton became the first licensed woman optometrist. In the late 1960s, there were only 368 women optometrists in the US, representing about 2% of practicing optometrists. Shortly after, the percentage of women in optometry schools grew to 19% in 1980 and even further to 44% by 1989. It was not until 1992 that women became majority of the students enrolled in optometry schools (Denial, 2019). Although there is still a greater number of practicing optometrists who identified as men compared to women, the percentage of practicing women continues to increase, and many more own practices now compared to a decade ago, 39.1% in 2016 compared to 20.5% in 2009 (*The future is female*, 2019).

Leadership

In the US in 1997, 53% of the optometric student body was women, yet 0% of deans and presidents of academic institutions reflected this demographic. According to the Association of Schools and Colleges of Optometry, the percentage of women optometry deans increased from 7% in 2002 to 29% in 2019. In 2022, nearly 70% of full-time Doctor of Optometry Students were women, and currently, 9 deans and 1 president of optometry schools are women. In 2022, the first woman Hispanic dean, Dr. Sandra Fortenberry, OD, FAAO, and the first black woman dean, Dr. Keshia Elder, OD, MS, FAAO, were appointed at the University of the Incarnate Word Rosenberg School of Optometry and the University of Missouri-St. Louis College of Optometry, respectively (Filkins & Maharjan, 2023).

The 2018 annual report of the American Academy of Optometry (AAO) noted that approximately 40% of fellows and approximately 63% of candidates for fellowship were women. The importance of role models is evident since Dr. Joan Exford became the first woman president of the AAO in 1993 she inspired more women to aspire for leadership positions. Subsequently, Dr. Karla Zadnik became the president of the AAO in 2011, followed by Dr. Barbara Caffrey in 2018. Dr. Dori Carlson became the first woman president of AOA in 2011, and Dr. Andrea Thau was elected to the role in 2016 (Filkins & Maharjan, 2023).

In Canada, the Canadian Association of Optometrists and the Canadian Ophthalmological Society have taken steps to increase diversity in leadership positions, including establishing Diversity and Inclusion Committees to promote EDI in the profession and provide resources for optometrists to better serve patients from diverse backgrounds (*Canadian Association of Optometrists*, 2024; *Canadian Ophthalmological Society*, 2024).

Similarly, in Europe and Australia, the British College of Optometrists established a Diversity and Inclusion Group, and the Royal Australian and New Zealand College of Ophthalmologists created a Diversity and Inclusion Committee to promote EDI in the profession (*General Optical Council: Equity and Diversity Data Monitoring Report*, 2021; Nirghin et al., 2011; *Optometry Board of Australia—Statistics*, 2023; *The optometrist and dispensing optician workforce in Aotearoa New Zealand*, 2023).

Overall, efforts to increase diversity in leadership roles in optometry, ophthalmology, and opticianry are important for promoting EDI in these professions and improving access to care for patients from underrepresented backgrounds (Auelua-Toomey & Roberts, 2022).

US longitudinal data other groups

There are limited longitudinal data available to quantify the growth of minority representation within the US optometry workforce. Limited historical student enrollment data do exist. A paper by Chen compiled optometry student data from 1969 to 2010, which revealed that Black students represented 0.63% of total enrollment in 1969 and was 2.71% in 2010; Hispanic students represented 1.00% in 1969 and was 4.52% in 2010, while Asian students increased from 3.43% in 1969 to 28.18% in 2010 (Chen, 2012). The number of minority students entering US optometry programs has continued to increase. Black students represented 4.3% of total enrollment in 2022; Hispanic students represented 8.7%, and Asian students represented 30.4% in 2022 (*Race/ethnicity of full-time doctor of optometry students* 2006–2023, 2023).

There are even less historical data regarding Native American or Asian American, Native Hawaiian/Alaska Native (AANHPI) groups within the optometric workforce or optometry student data. In general, the classification and reporting of these groups have been less standardized, and given the small population sizes, data from multiple groups are often aggregated (Wang Kong et al., 2022). For example, Chen combined all students who were not reported as White, Black, Hispanic, or Asian into the category of "Other" because student data was inconsistent over time (Chen, 2012). In 1969, the percentage of students in this aggregated group was 0.01%, which grew to 0.08% by 2010 (Marshall, 1972).

The 2020 US Census indicated that the US population was becoming more multiracial over time. This population was 9 million people in 2010 and grew to 33.8 million people in 2020, a 276% increase (Nicholas Jones et al., 2021). Currently, there are no data on the number of optometrists who identify as two or more races. The Associated Schools and Colleges of Optometry began collecting information on students who identify as two or more races in 2015. At that time, the percentage of students who identified with this group was 2.1% and grew to 2.9% by 2022 (Marshall, 1972).

Rationale for systemic action, accompanied by solutions, interventions, and possible programs/initiatives to tackle disparities

Pipeline programs

Optometry is a profession that many students are unaware of as a career option in healthcare. Traditionally, it is promoted to undergraduate students through prehealth advisors, undergraduate clubs, and practicing optometrists. However, the majority of practicing optometrists do not identify with historically marginalized groups, which adds to the difficulty of advocating for the profession with prospective students from those groups. Despite the challenges at hand, one avenue through which optometry can be promoted is through pipeline programs that introduce students to the discipline at a younger age. These pipeline programs present optometry as an option within the healthcare professions for students exploring their future career options.

Pipeline programs expose students to the discipline and offer practicum opportunities and potential for mentorship. Students report the positive impact of role models and mentorship on their ability to achieve their goals, with optometry being no exception. Many pipeline programs exist in undergraduate institutions or are available to undergraduate and graduate-level students. These programs serve to introduce the profession of optometry to underrepresented minority students by allowing participants to interact with current optometric students, shadow classes and labs, and even assist students with navigating the OAT preparation and building their portfolios. There are many schools of optometry that offer programs, such as the Collegiate Science and Technology Entry Program at SUNY College of Optometry and the Summer

Enrichment Program at the Pennsylvania College of Optometry at Salus University (*Optometry Summer Programs—ASCO*, 2024). Both of these programs target undergraduate and graduate-level students.

While programs, such as those listed target college and graduate-level students, there is still an opportunity for concerted efforts aimed at targeting younger students. Middle and high school students are often exploring opportunities within their schools and districts. Additionally, many students in this age range seek work experience and have an interest in internships. Providing such opportunities in the optometric discipline would allow for earlier exposure and greater opportunities for future study and career. Programs focused on younger students may not be as common due to the lack of an immediate return on investment; with years of education to complete before entering optometric programs of study, some could lose interest and be diverted onto different paths. Optometric schools and programs must understand the marathon versus the sprint; while return on investment may not be seen for years or even decades, the foundation built in that time will allow for further reaching, stronger and deeper connections with young and diverse communities.

EDI training

There are tangible benefits of EDI including research that teams with diverse perspectives have enhanced problem-solving capabilities. The adoption of EDI into the business's culture has been described as good for business. EDI results in an improved employee morale, reduced burnout, improved employee retention and engagement, and increased profitability and success (Ding et al., 2023; Rosales et al., 2023). The concept of diversity in healthcare is relatively new and is gaining recognition and importance. Concerning healthcare specifically, promoting EDI can help create an environment of psychological safety and a sense of belonging, and increase workforce well-being (Yong-Hing et al., 2023; Young et al., 2023). However, it is important to note that there are additional important metrics of value in healthcare that depend on the perspectives of stakeholders, including patients, learners, and researchers. One of the most essential values of diversity in healthcare is cultural humility, provision of culturally competent care, and the resulting enhancement of patient satisfaction and better outcomes (Yong-Hing & Khosa, 2023).

While exploring pipeline and pathways programs, we must also confront the existing disparities in education. Future planning, such as pipeline and pathway programs, must be coupled with current strategies to increase education, awareness, and strategy to confront structural inequities within our practices, schools, and organizations. These inequities often mirror those in the societies we work and live in. Factoring in the historical imbalance of those leading and working within the optometric discipline lends further substantiation of the disparities that exist today, and the need for training and resources to combat the biases that may subsequently permeate the discipline.

As is true in many professions, there is a paucity of literature on EDI and most of the available information offers little remedy to the intractable problems. Many lack awareness of the limitations on their knowledge due to perhaps holding privilege within their identities and lived experiences. As such, there is a lack of progress in the learning and unlearning processes, for we often are unaware of the gaps in our knowledge until they are made visible to us. This cloaking of truth and learning is perpetuated by our educational systems at large, which politicize the teaching of accurate history; many subjects and teaching methods come under fire for a perceived threat of "EDI Indoctrination" (*Critical race theory in schools isn't indoctrination, it's the truth*, 2022). This is evident in the recent standards approved by the Florida Board of Education that limits courses, such as a pilot-advanced placement high school course on African American studies. The standards also present guidelines on how to teach Black history, thus impeding antiracist efforts in Floridian schools and curricula (Fortinsky, 2023). Such politicization of the educational system contributes largely to a divided and segregated society.

When education and training does not equip people with the vocabulary to describe themselves and others, their ability to understand similarities and differences is hampered. While diversification efforts are critical, we cannot bring more diverse populations into environments in which they may not be set up to succeed due to a lack of resources, understanding, and policies that are not designed for such diversity. Educational opportunities and efforts to identify and address current inequities must couple with our future diversification, for, if we do not disrupt the pattern, we will be condemned to repeat it.

The differences that we all hold in our intersectional identities, along with our different lived experiences, teach us the lens through which we see the world and act within it: with each other, with our patients, and in our professional endeavors. Training programs implemented at all levels for students, faculty, and staff within schools and programs and continuing education for practitioners are vital steps forward.

Education is key in all areas, and intentional and diverse methods of instruction are critical. While training programs exist in most organizations, there must be opportunities for group and individual learning, safe spaces for questions and connection, and content that is specific and tailored to the group to avoid the common pitfalls of EDI training. Examples of such pitfalls include a lack of engagement, lack of interest, and at times, harm done to participants when training is poorly or inexpertly delivered. Timing and frequency are also important, as many training programs fail to answer all questions or prepare participants in how they can make immediate changes. To address this, programs must provide ongoing opportunities for learning both to provide repetition where necessary, and to provide first-time learning opportunities for new community members. Additionally, it is beneficial to provide scaled learning options, so participants can continue to advance and practice skills while confronting new content and topics.

While topic and content-driven opportunities are important, it is also important to build in opportunities for connection and relationship building. Much of this work is individual, but before and even during individual learning, communities must have the opportunity to work and learn together. Most EDI training stresses the importance of seeing outside of one's own perspective, and to do so, we must explore avenues through which we can learn about each other. This approach is often ignored or abandoned due to perceived difficulty in execution, or a perception that such opportunities are not as valuable as structured trainings even though many organizations benefit from such programs.

Providing training and educational programs together with pipeline programs will allow for a twofold effect: diversification of the future of healthcare professions while we also educate and prepare ourselves to better serve our communities, patients, and students of today (Yong-Hing & Khosa, 2023). Change will not come immediately, but by considering our current and future impacts, we can more effectively strive toward our goals and tackle the disparities facing our communities and healthcare professions (Purnell et al., 2016).

Responsibilities of governing groups and industry

The death of George Floyd in the summer of 2020 and Asian Americans and Pacific Islanders hate in 2021 was an awakening not only for the world but also for all the healthcare professions. All key associations in optometry (American Academy of Optometry, American Optometric Association, Association of Schools and Colleges of Optometry) and the ophthalmic industry began to reflect and look inward to the necessity of EDI. Each leadership body created or revised their EDI statements and policies to comment on intolerance, bias, racism, discrimination, and the importance of inclusion and representation.

Numerous EDI committees were formed, special interest groups were invigorated, strategic plans were developed, and money and resources were allocated to pipeline programs and scholarships. In addition, employee training programs, podcasts, industry news articles, research, continuing education lectures, and optometric curricula were reviewed and developed.

A myriad of research articles have clearly shown that a diverse workforce provides new ideas and is good for business. However, there is a duty with the collective leadership of the optometric profession and industry to go beyond diversity and inclusion. In reviewing the various EDI statements and policies, there is a clear focus on diversity and inclusion, but lacks follow-up to move into a direction that ensures equity and justice.

The American Medical Association has called out past wrongs, examined systemic and institutionalized racism, and developed a robust strategic plan that fosters pathways for truth, healing, reconciliation, and transformation, now and in the future (*The AMA's strategic plan to embed racial justice and advance health equity*, 2024). Optometry on the other hand is in the beginning of this journey, and more data need to be collected to examine its blind spots that affect the pipeline for the profession, current optometrists, and the industry it serves.

A strategy to embed justice and health equity will aid the profession to emerge as one that will aid current optometrists, future optometrists, and the patients and communities we serve. Optometry requires this fortitude to move and fully embrace the ideals of EDI and justice.

Conclusion

As we continue to live in challenging times, in many respects, the issues and opportunities of today are not new. It was in 1972 when Marshall wrote, "It is appalling to think it took over 300 years for the souls of the masses of this great nation to be touched by the social concern that is so prevalent today." What may appear like new conversations are essentially extensions of more than five decades of dialogue along the justice and EDI journey (Marshall, 2022).

"Marathon" is often cited as a metaphor for the EDI work that lies before us but let us not lose sight of the fact that the optometry diversity marathon started decades ago and continues today as a slow race to progress. It is beyond time to push forward in solidarity with enhanced mindfulness and humility, socially conscious and just values, clearly focused and purposeful intention, and transformative action to achieve a profession with a racial, ethnic, and cultural complexion that mirrors the population it intends to serve and one that is prepared to advance equitable care and remedy the vision and eye health disparities so prevalent among racialized communities (Marshall, 2022).

We have not achieved our goals, but this has been the impetus for where we are today. Looking at the past and the present with honesty helps us plan and dream for the future. This will help the profession and the individuals and communities we serve. Because of the past, our leadership is not diverse. While there is increasing women representation, diversity is lacking in the dimensions of race, ethnicity, religion, sexual identity, and many others.

To increase diversity in the profession, it always starts with students. Through this effort and journey, future leaders will be developed to help achieve the goals of EDI for the profession and the patients we serve.

These efforts will ensure a wide range of experiences and voices are represented in all sectors of care. The demographics of our communities are rapidly changing, so it is crucial to equip our healthcare providers with the cultural competency and awareness needed to deliver quality care to diverse populations. This approach benefits practice revenue, patient outcomes, and the overall well-being of the communities we live and work in Gomez and Bernet (2019).

We can continue to diversify and expand care within the optometric discipline in several ways. Data collection is at the forefront of dynamic change as it will be able to provide a clearer picture of current affairs and highlight areas for improvement as we continue to progress. This is the baseline for many of the programs that champion EDI efforts. Those within the eyecare industry need to collaborate to support and create more pipeline programs that will provide diversity and improve care within rural communities. These programs need to be supported by a two-pronged approach: early outreach to STEM (Science, Technology, Engineering, and Math), students to inspire careers in optometry and mentorship programs that will help foster success during and after their time in optometry school (Toretsky et al., 2018).

Individual and tailored resources are essential since every person is unique. There cannot be a one-size-fits-all approach to these programs. Long-term change can only be achieved if collective and sustained efforts exist throughout the healthcare professions. One-sided, temporary efforts will not bring the level of improvement that is needed. Contextualized, data-driven approaches are necessary to move forward (Toretsky et al., 2018). All patients deserve to be able to communicate and trust their healthcare providers, whether they are being seen for routine exams or managing complex conditions. EDI efforts are not simply another box to check but are imperative to providing patient-centered care. With this, optometrists can safeguard all individuals' vision quality and ocular health, regardless of background.

Call to action

1. To increase diversity and the number of racially minoritized students who graduate from the schools and colleges of optometry, action is required. We recommend the following:
   - Collection of data regarding all aspects of education (diversity of applicants, current students, residents, and faculty); without a complete picture of the current status, it will be difficult to measure progress and eventual success.
   - Track reasons why racially minoritized students drop out of optometry school.
     o Underperformance is a complex issue that needs to be carefully studied and addressed.
     o Mentoring, counseling, and financial investment are needed for racially minoritized students who may not have the resources to help them succeed in a rigorous academic program; various cultural differences and needs exist, which will necessitate individualized support and resources to appropriately meet the needs of these students.
   - Collectively pool resources to develop pipeline programs for racially minoritized students and not compete with one another.
   - Cross-pollinate successes from STEM programs and other health professions.
   - Extensive strategy to promote the profession of optometry to communities who traditionally have little understanding of optometry.
   - Market the profession beyond undergraduate programs to include middle schools and high schools.

- Understand the return on investment is not immediate (it may take a decade, but without investing with a clear and committed strategy, we will be in the same position 10 years from now).
2. To improve equity, inclusion, and justice in the profession of optometry, action is required. We recommend the following:
   - Develop a robust strategic plan that acknowledges past wrongs and moves to a direction of embedding justice and health equity.
   - Required EDI training in statutes and regulations of optometry for each governmental jurisdiction.
   - Include discussions of social determinants of health in continuing education with examples of racially minoritized groups and communities that are involved.
   - Require EDI training and criteria in the accreditation process of colleges and schools of optometry.

EDI statements/policies

ARBO—Association of Regulatory Boards of Optometry
ARBO's statement regarding diversity, equity, and inclusion
The Association of Regulatory Boards of Optometry (ARBO) recognizes the existence of intolerance, exclusion, racism, and other forms of discrimination and bias. ARBO is committed to promoting equity and justice in all its business and regulatory practices and policies.

AAO—American Academy of Optometry
Diversity, equity, inclusion, and belonging statement
The American Academy of Optometry (The Academy) promotes diversity, equity, inclusion, and belonging in all aspects of our work. We recognize that our volunteer members play a critical role in shaping the future of our organization, and we are dedicated to ensuring that they represent our membership and the patients we serve.

We believe that a diverse organization is a better organization. By bringing together individuals with a wide range of backgrounds, experiences, and perspectives, we can more effectively identify and address the needs of our membership. We know that having different viewpoints helps generate better ideas. In addition to knowledge, skills, and experience, consideration is given to appointing Academy Fellows who have not previously served on a committee (or task force) and recruiting members who reflect the full spectrum of our membership, including but not limited to diversity in terms of race, ethnicity, gender, sexual orientation, disability, age, mode of practice, geographic location, and area of optometric specialty.

The Academy is also committed to creating an inclusive and welcoming environment for all committee members, where everyone feels valued and respected. We are dedicated to promoting open and respectful communication and creating opportunities for all volunteer members to contribute and succeed.

In short, the more inclusive we are, the better our work will be. Diversity is not only the right thing to do but also essential to our success as an organization.

ASCO—Association of Schools and Colleges of Optometry

ASCO's strategic plan addresses challenges to diversity in optometric education and identifies three main goals:

1. Build awareness of the profession of optometry as a highly valued and rewarding career choice among **prospective students** and their influencers (namely, their parents and career advisors).
2. Provide resources, support, and encouragement to **faculty and administrators** in building an inclusive organizational culture that welcomes diverse students who will feel like they belong in both their educational institutions and in the field of optometry.
3. Support culturally competent training that enables **students** to provide top-quality care to an increasingly diverse population of patients.

AOA—American Optometric Association

AOA's commitment to diversity and inclusion

The American Optometric Association upholds and advances diversity, equity, and inclusion. AOA is committed to advancing the diversity of the profession of optometry, nurturing an organizational culture that is committed to diversity and inclusion and continuing efforts to ensure access to care. These endeavors enable AOA to further the profession of optometry and improve the eye, vision, and overall health of the public.

Enhancing the diversity of the profession of optometry

We are committed to implementing initiatives that attract more underrepresented minorities to the profession; engage diverse doctors, students, and paraprofessionals in leadership roles, and provide the tools and resources to support them, including continuing education.

Nurturing an organizational culture committed to diversity and inclusion

AOA is committed to nurturing a staff culture that embraces the value of diversity and inclusion through ongoing training and collective efforts aimed at reinforcing the importance of a diverse and inclusive team. Advancing the principles of diversity and inclusion in the workplace is important to improved collaboration and morale, as well as greater innovation, productivity, and representation in the work we do to uphold the profession of optometry and public access to the care Doctor of Optometry deliver to the public.

Disclosures

All authors have no financial disclosures.

References

A road map for diversity in optometric education and the profession. (2004). Association of Schools and Colleges of Optometry.

Auelua-Toomey, S. L., & Roberts, S. O. (2022). The effects of editorial-board diversity on race scholars and their scholarship: A field experiment. *Perspectives on Psychological Science: A Journal of the Association for Psychological Science, 17*(6), 1766–1777. https://doi.org/10.1177/17456916211072851.

Board of Directors—World Council of Optometry. (2024). World Council of Optometry. https://worldcounciloptometry.info/leadership/.

(2024). *Canadian Association of Optometrists.* https://opto.ca/.

Canadian Ophthalmological Society. (2024). COS-SCO. https://www.cos-sco.ca/.

Chen, E. E. (2012). Comparing proportional estimates of U.S. optometrists by race and ethnicity with population census data. *Optometric Education, 37*(3). https://journal.opted.org/articles/Volume37_number3_Article1.pdf.

Cole, J. (2021, July 15). Optometry at work: The how, when and where of who delivers care. *Review of Optometry.* https://www.reviewofoptometry.com/article/optometry-at-work-the-how-when-and-where-of-who-delivers-care.

Critical race theory in schools isn't indoctrination, it's the truth. (2022, January 24). New University. https://newuniversity.org/2022/01/24/critical-race-theory-in-schools-isnt-indoctrination-its-the-truth/.

Denial, A. (2019). A look back: Celebrating women in optometry. *Optometric Education.* https://journal.opted.org/article/a-look-back-celebrating-women-in-optometry/.

Ding, J., Yong-Hing, C. J., Patlas, M. N., & Khosa, F. (2023). Equity, diversity, and inclusion: Calling, career, or chore? *Canadian Association of Radiologists Journal, 74*(1), 10–11. https://doi.org/10.1177/08465371221108633.

Elder, K. S. (2022). On diversity in optometry, progress, but more work to be done. *Optometric Education.* https://journal.opted.org/article/on-diversity-in-optometry-progress-but-more-work-to-be-done/.

Filkins, K. J., & Maharjan, E. K. (2023, April 20). Eyes on the future: Meet the female deans leading optometric education. *Optometry Times.* https://www.optometrytimes.com/view/eyes-on-the-future-meet-the-female-deans-leading-optometric-education.

Fortinsky, S. (2023, July 19). Florida Board of Education approves controversial standards for teaching black history. *The Hill.* https://thehill.com/policy/4106951-florida-board-of-education-approves-controversial-standards-for-teaching-black-history/.

General Optical Council: Equity and Diversity Data Monitoring Report. (2021). General Optical Council. https://optical.org/media/mgqjuo1c/equality-and-diversity-data-monitoring-report-2021.pdf.

Gomez, L. E., & Bernet, P. (2019). Diversity improves performance and outcomes. *Journal of the National Medical Association, 111*(4), 383–392. https://doi.org/10.1016/j.jnma.2019.01.006.

Harewood, J., & Rosenfield, M. (2021). Defining race and ethnicity in optometry. *Ophthalmic and Physiological Optics, 41*(4), 659–662. https://doi.org/10.1111/opo.12846.

Jones, N., Marks, R., Ramirez, R., & Rios-Vargas, M. (2021, August 12). *2020 Census illuminates racial and ethnic composition of the country.* United States Census Bureau. https://www.census.

gov/library/stories/2021/08/improved-race-ethnicity-measures-reveal-united-states-population-much-more-multiracial.html.

Kuehn, B. M. (2021). Preventable vision loss affects millions globally. *JAMA, 325*(15), 1498. https://doi.org/10.1001/jama.2021.4494.

Marshall, E. C. (1972). Social indifference of blatant ignorance. *Journal of the American Optometric Association, 43*(12), 1261–1266.

Marshall, E. C. (2022). Justice and disparity—A defining cause for diversity, equity and inclusion in optometric education and practice. *The Journal of Optometric Education. Optometric Education.* https://journal.opted.org/article/justice-and-disparity-a-defining-cause-for-diversity-equity-and-inclusion-in-optometric-education-and-practice/.

(2021, January 7). *Massachusetts scope win adds glaucoma authority.* https://www.aoa.org/news/advocacy/state-advocacy/massachusetts-scope-win-adds-glaucoma-authority?sso=y.

McMorris, E. (2016). Intro to the health workforce in Canada: Optometry. *Canadian Health Workforce Network..* https://www.hhr-rhs.ca/images/Intro_to_the_Health_Workforce_in_Canada_Chapters/15_Optometry.pdf.

Nirghin, U., Khan, N. E., & Mashige, K. P. (2011). Institutional, gender and racial profiles of South African optometrists. *African Vision and Eye Health, 70*(3), 123–128. https://doi.org/10.4102/aveh.v70i3.107.

Optometry Board of Australia—Statistics. (2023). Optometry Board. https://www.optometryboard.gov.au/About/Statistics.aspx.

Optometry Summer Programs—ASCO. (2024). Association of Schools and Colleges of Optometry. https://optometriceducation.org/diversity/optometry-summer-programs/.

Optometry's reflection. (2021, April 7). American Optometric Association. https://www.aoa.org/news/inside-optometry/aoa-news/diversity-optometrys-reflection?sso=y.

Purnell, T. S., Calhoun, E. A., Golden, S. H., Halladay, J. R., Krok-Schoen, J. L., Appelhans, B. M., & Cooper, L. A. (2016). Achieving health equity: Closing the gaps in health care disparities, interventions, and research. *Health Affairs, 35*(8), 1410–1415. https://doi.org/10.1377/hlthaff.2016.0158.

Race/ethnicity of full-time doctor of optometry students 2006-2023. (2023). U.S. Schools and Colleges of Optometry. https://optometriceducation.org/wp-content/uploads/2023/05/Enrollment-Race-1-1.pdf.

Rosales, R., León, I. A., & León-Fuentes, A. L. (2023). Recommendations for recruitment and retention of a diverse workforce: A report from the field. *Behavior Analysis in Practice, 16* (1), 346–361. https://doi.org/10.1007/s40617-022-00747-z.

Shneor, E., Isaacson, M., & Gordon-Shaag, A. (2021). The number of optometrists is inversely correlated with blindness in OECD countries. *Ophthalmic and Physiological Optics, 41*(1), 198–201. https://doi.org/10.1111/opo.12746.

The AMA's strategic plan to embed racial justice and advance health equity. (2024, April 9). American Medical Association. https://www.ama-assn.org/about/leadership/ama-s-strategic-plan-embed-racial-justice-and-advance-health-equity.

The future is female. (2019, March 7). American Optometric Association. https://www.aoa.org/news/inside-optometry/aoa-news/the-future-is-female?sso=y.

The optometrist and dispensing optician workforce in Aotearoa New Zealand. (2023). Optometrists and Dispensing Opticians Board. https://www.odob.health.nz/document/8129/ODOB%20Workforce%20in%20ANZ%202024%20.

Toretsky, C., Mutha, S., & Coffman, J. (2018, July 30). *Breaking barriers for underrepresented minorities in the health professions.* Healthforce Center at UCSF. https://healthforce.ucsf.edu/publications/breaking-barriers-underrepresented-minorities-health-professions.

Twa, M. D. (2020). An editor's journey toward diversity and equity. *Optometry and Vision Science*, *97*(7), 471–472. https://doi.org/10.1097/OPX.0000000000001549.

Wang Kong, C., Green, J., Hamity, C., & Jackson, A. (2022). Health disparity measurement among Asian American, native Hawaiian, and Pacific Islander populations across the United States. *Health Equity*, *6*(1), 533–539. https://doi.org/10.1089/heq.2022.0051.

Yashadhana, A., Clarke, N. A., Zhang, J. H., Ahmad, J., Mdala, S., Morjaria, P., Yoshizaki, M., Kyari, F., Burton, M. J., & Ramke, J. (2021). Gender and ethnic diversity in global ophthalmology and optometry association leadership: A time for change. *Ophthalmic and Physiological Optics*, *41*(3), 623–629. https://doi.org/10.1111/opo.12793.

Yong-Hing, C. J., & Khosa, F. (2023). Provision of culturally competent healthcare to address healthcare disparities. *Canadian Association of Radiologists Journal*, *74*(3), 483–484. https://doi.org/10.1177/08465371231154231.

Yong-Hing, C. J., Vaqar, M., Sahi, Q., & Khosa, F. (2023). Burnout: Turning a crisis into an opportunity. *Canadian Association of Radiologists Journal*, *74*(1), 16–17. https://doi.org/10.1177/08465371221130683.

Young, P. J., Kagetsu, N. J., Tomblinson, C. M., Snyder, E. J., Church, A. L., Mercado, C. L., Guzman Perez-Carrillo, G. J., Jha, P., Guerrero-Calderon, J. D., Jaswal, S., Khosa, F., & Deitte, L. A. (2023). The intersection of diversity and well-being. *Academic Radiology*, *30*(9), 2031–2036. https://doi.org/10.1016/j.acra.2023.01.028.

Chapter 8

Equity, diversity, and inclusion in pharmacy: Paramount to progress and public trust in the profession

Sunny Bhakta[a], Mahreen Khosa[b], Richard Ogden Jr.[c], and Anthony Scott[d]

[a]*Department of Pharmacy Services, Houston Methodist Hospital, Houston, TX, United States,* [b]*Department of Pharmacy, Mater Private Hospital, Dublin, Ireland,* [c]*Department of Pharmacy, Children's Mercy Kansas City, Kansas City, MO, United States,* [d]*Emory University Hospital, Atlanta, GA, United States*

We will all profit from a more diverse, inclusive society, understanding, accommodating, even celebrating our differences, while pulling together for the common good.

—Ruth Bader Ginsburg

Promoting equity, diversity, and inclusion (EDI) in the pharmacy profession is crucial for ensuring equitable healthcare delivery both in the United States and internationally. This chapter underscores the significance of fostering a diverse workforce to reflect the care offered to various patient populations. Achieving EDI in pharmacy involves addressing disparities in recruitment, education, and career advancement. Emphasizing formalized training and educational competence in EDI-specific topics is essential for advancing the profession and providing patient-centered care. Moreover, the chapter stresses the importance of creating an inclusive workplace culture that values different perspectives and backgrounds. EDI initiatives not only enhance the profession's representation but also contribute to improved outcomes by acknowledging and addressing diverse populations. In summary, cultivating EDI in the discipline of pharmacy is integral to advancing the pharmacy workforce, achieving equitable outcomes, and fostering a more responsive and all-encompassing healthcare environment.

Pharmacists remain the most accessible healthcare professionals for people from all walks of life around the globe. The profession of pharmacy is not exempt from concerns surrounding EDI. Demographically, the pharmacy

Equity, Diversity, and Inclusion in Healthcare. https://doi.org/10.1016/B978-0-443-13251-3.00008-9

profession in Europe and North America appears diverse; however, more efforts must be made to address gender and racial disparities in leadership positions. Owing to the unique placement of pharmacists at the heart of communities, a strategic focus on addressing healthcare needs of diversifying communities is needed to garner trust and improve health outcomes of the population at large.

The pharmacy profession is made up of several different sectors that have unique attributes. These include health system, community, ambulatory, academia, regulatory/managed care, industry, and research. Health system pharmacies include hospitals (nonprofit or investor-owned) and long-term care facilities. In North America, these serve as traditional sites for postgraduate residency training. These pharmacy locations are generally clinical in training and practice and have interactions with many different patient populations and other healthcare disciplines. Often, health system pharmacists are focused on acute or episodic care of patients. Health systems can be nonprofit or private.

Community pharmacies can be independently owned or part of a local/national chain. They can also be freestanding (mostly freestanding in Europe) or embedded pharmacies within other retail organization (retailer, grocery store, etc.). In North America, community pharmacies have more recently served as sites of postgraduate residency training; however, in Europe, pharmacy students are typically trained in community pharmacies. It is important to note that postgraduate training in most European countries is informal. Community pharmacists have frequent interactions with patients in one community and more opportunities for creating longitudinal relationships within their community. When compared to health system pharmacists, community pharmacists have less involvement with other healthcare disciplines, typically interact solely with prescribers (doctors and in some jurisdictions nurses) in relation to prescription queries, and traditionally have an increased involvement in the financial side of the pharmacy business.

Ambulatory pharmacy practice is primarily based on physician practices with a focus on patients with chronic conditions. Pharmacists in this setting may participate in collaborative practice or protocol-driven agreements and currently are an emerging site for postgraduate training. Academia pharmacy practice is either full- or part-time focused in college or university settings and may instruct in other healthcare disciplines (medicine, nursing, dental, respiratory therapy, etc.). Depending upon their site and scope of practice, they may have variable access to direct patient care.

Managed care pharmacists expand from traditional pharmacy sectors into the payor domain. These include pharmacy benefits management, comprehensive medical insurance, and group medical insurance. The focus of this sector is less on the individual patient and more on population health. In the Republic of Ireland and other European countries, while there is no such defined role for pharmacists in the health sector(s), pharmacists are involved in government departments focusing on cost-effectiveness analysis of new drugs coming to the market, drug pricing negotiation, reimbursement on national drug access schemes, and selection of preferred drugs at a national level. This focus on cost versus benefit analysis reduces the cost for health systems and functions to reduce disparities that exist due to the financial ability of cohorts within populations to access healthcare.

The pharmaceutical industry can be clinical, operational, research, or sales-focused. Much like academia, depending upon their site and scope of practice, they may have variable access to direct patient care. Interaction with other healthcare disciplines depends on the role of the organization. Many of these practitioners will work directly for a pharmaceutical manufacturer, so they will have a greater emphasis on the innovation and business side of healthcare. Pharmacists who practice in research settings focus less on directly treating patients and more on new or novel therapies for a controlled patient population. These may be freestanding research facilities or affiliated with an academic institution. As clinical trials have varying levels of direct patient interactions, pharmacists in research roles may have variable or no interactions with patients.

Equity, diversity, and inclusion in healthcare: Current outlook, trends, and future perspectives

EDI are essential principles both in healthcare and the pharmacy profession. These pillars signify a culture that promotes equitable access to training and career opportunities for all, as well as care for patients, enhanced patient outcomes, and a culturally competent workforce that reflects the community it serves. It is essential to provide a comprehensive analysis of the current state of EDI within healthcare and pharmacy. This should include highlighting published statistics, critical reviews of current workforce statistics, noting improvements over the past few years, as well as setbacks, stagnation, and future trends within the category.

Owing to the diversity of the demographic composition of both the United States (US) and European populations, pharmacists in today's world can work closely with patients of different racial/ethnic backgrounds and other underrepresented categories of inclusion. The 2020 socio-political landscape in the US during the height of the COVID-19 pandemic made EDI visibility and action items a priority for many organizations and institutions with pharmacy representation. The structural element of EDI education and awareness in pharmacy is largely comprised of, but not limited to, academia, professional organizations, healthcare systems, and pharmaceutical companies (George et al., 2023).

As a baseline, the published statistics related to EDI in pharmacy reflect the need to prioritize these initiatives. For example, studies show that gender, racial and ethnic disparities are prevalent in executive positions across academic disciplines and healthcare sectors. While reports indicate the profession has become increasingly diverse, the numbers still shed light on the lack of hires from underrepresented populations within the healthcare workforce. The racial diversity reviews show that the percentage of non-White licensed pharmacists across the US has increased to 21.8% between 2014 and 2019, which is a 46% increase in this period. In the United Kingdom (UK), 48.1% of pharmacists identify as Black, Asian, and minority ethnic groups (Royal Pharmaceutical Society, 2023). These figures may be attributed to the UK's extensive colonial history and the immigration of individuals from these regions into the UK.

Freedom of movement is one of the fundamental cornerstones of the European Union and has driven European integration since the 1950s. This means that professional qualifications from one member state are readily recognized by others. This promotes diversity in European workforces, as evidenced by figures from the workforce intelligence report commissioned by the Pharmaceutical Society of Ireland (PSI), demonstrating that 49% of pharmacists in the workforce were qualified outside of the Republic of Ireland (Pharmaceutical Society of Ireland, 2023a). However, an area that requires additional work is that of third country qualification recognition (TCQR), which would allow pharmacists qualified outside of the European Economic Area to have their qualifications recognized and subsequently be allowed to register and practice as pharmacists in the European Union (Pharmaceutical Society of Ireland, 2023b). It is currently a lengthy and multistage process. To combat ongoing capacity issues in the Republic of Ireland, the PSI has undertaken a project to streamline the TCQR route. This project has taken inspiration from Canadian practices on the issue. In Canada, internationally educated pharmacists are required to undertake a 4–6-month university-based bridging program, which is focused on equipping them with the necessary knowledge and skills of pharmacy practice and therapeutics required to practice in Canada. This is coupled with a 6–12-month in-service and clinically based training component and pharmacy-specific English language courses, which have been developed to help pharmacists meet the language requirement to practice in any setting. The development of programs such as this one promotes EDI by ensuring equality of access to professional opportunities for pharmacists of different ethnic and racial backgrounds.

The path to diversity begins with supporting, mentoring, and sponsoring diverse women and men to become leaders and entrepreneurs.

—Denise Morrison

Underrepresentation of diverse racial and ethnic backgrounds can also be seen in executive positions across healthcare sectors. This is an area of focus for development for the Royal Pharmaceutical Society (RPS) in the UK, as evidenced by the pharmacy workforce race equality standard report of 2023 (National Health Service, 2023a). Hindrances to the development of action plans include gaps in the data that make it difficult to assess the other categorizations of underrepresented populations in a way that can create purposeful action items and goals. Reports such as this one provide valuable data on the demographic composition of the pharmacy profession and opinions on the state of EDI in the profession currently. It is therefore important that more reports of this nature are commissioned and reviewed by pharmacy regulators and interest groups across the globe.

The prevalent gender disparity in academic disciplines and healthcare professions also pervades pharmacy and leadership positions, as indicated by the diversification metrics. The International Pharmaceutical Federation (FIP) predicts that by the year 2030, more than 70% of the global pharmacy workforce will be comprised of women (Bukhari et al., 2020). However, women are underrepresented in executive roles and decision-making positions, hindering the achievement of gender equity within healthcare. Across the globe, studies have revealed that pharmacy as a profession continues to be dominated by women, with nearly two-thirds of the workforce comprised of women compared to only 46.4% in 2009. The upward trend continues for women in leadership roles; 58.8% of pharmacists in management positions in 2019 were women (Martin et al., 2021). These statistics reflect the education enrollment gaps between women and men in pharmacy school, with women comprising over 60% of academic students and graduates in the US. Published data on the inclusive representation of applicants and enrollees that identify as nonbinary or nonconforming is a current gap in the available data (Data USA, 2019; Tanni & Qian, 2021). In Ireland, 65% of pharmacists were women, and 35% of pharmacists were men. Interestingly, this report detailed that of those pharmacists involved in leadership roles, 55% were women and 45% were men (Pharmaceutical Society of Ireland, 2023a).

In addition to significant findings in racial and gender diversity, studies find that the pharmacy profession is continuing to trend younger in its composition. Over 47% of practicing pharmacists (47.7%) were aged 40 years or younger in 2019, compared to just 24.4% for the same age demographic in 2009. According to the study authors, this is likely due to recent growth in pharmacy graduates and the simultaneous retirement of older pharmacists. The demographics of the pharmacy technician profiles in the US are not much different. As of 2020, there were over 380,000 licensed pharmacy technicians currently employed in the US. Overall, 73% of all licensed technicians are women, and only 26.3% are men. The average licensed pharmacy technician is 38 years old, and the most common ethnicity of licensed pharmacist technicians is White (60.2%), followed by Hispanic or Latino (14.5%), Asian (10.1%), and Black or African American (8.6%). In the Republic of Ireland, the average age of an Irish pharmacist according to the PSI workforce intelligence report in 2022 was 40.25 years (Pharmaceutical Society of Ireland, 2023a). Similar data have not been collected for pharmacy technicians. Studies also report earnings gaps between women and men in the profession (Data USA, 2019).

To gain deeper insights into the experiences and challenges faced by minorities and underrepresented populations, reviews, and publications such as those from the *American Journal of Health-System Pharmacy* (*AJHP*) on EDI provide valuable perspectives. A good starting point is to examine the barriers and facilitators of recruiting and retaining diverse healthcare professionals, with a particular focus on underrepresented minorities (e.g., age, race, gender identities). There are several pharmacy-specific initiatives aimed at increasing diversity in the workforce throughout the educational pipeline and through other means of development (Scott & Ogden, 2022). A joint initiative launched by the RPS, the association of pharmacy technicians in the UK, and 13 other national partner organizations titled "Inclusive Pharmacy Practice" is focused on making the workplace more inclusive for pharmacy professionals. One of the aims is to develop senior leadership in the profession that reflects the diversity of the communities it serves and the body of professionals within it, as this will undoubtedly improve health inequalities within the population (National Health Service, 2023b). The RPS recommends the introduction of initiatives that look to address the disparity between grades achieved in the program by ethnic minorities in comparison to their White pharmacy student counterparts. Workforce diversity efforts in the academic realm have focused primarily on structural diversity (such as representation of marginalized and underrepresented groups) among learners, faculty, and leaders of the profession (Crews et al., 2022). There are also psychological climates to consider (such as reported experiences of discrimination) and behavioral dimensions (such as mentorship). There is also a significant impact on EDI in the workforce based on the geographic locations of pharmacists, as reflected within the different states in the US and internationally (Smith et al., 2021).

Another area of focus is removing bias, whether conscious or subconscious, in fitness to practice (FTP) procedures. This was suggested following findings that non-White ethnicities were found to be overrepresented in referral to FTP procedures in the UK (National Health Service, 2023b). There was also a statistically significant relationship found between concerns raised being investigated against Asian pharmacists compared to all other races, with 46% of claims for FTP being associated with Asian pharmacists and 52% of these concerns going on to investigation. There was a suggestion that these referrals to FTP and indeed investigation of the same may have been rooted in bias against non-White pharmacists.

In the international context, the International Pharmaceutical Federation (FIP) set equity and equality development goals for pharmacy practice around the world. One such program is the National Alliance for Women in Pharmacy (NAWP) Initiative in Pakistan (International Pharmaceutical Federation, 2019). This initiative comes from the context of the discovery that most of the pharmacy professional leadership in Pakistan were men. It aims to raise awareness regarding gender equity and promote leadership development for women by offering organizational support, mentoring opportunities, networking platforms, and engagement outlets for women. It has also identified a key role in initiating policy decisions and awareness regarding security at the workplace, pay gaps, and safe working environment. Through NAWP, the concept of gender equity and creating an enabling environment for women and women leadership are discussed for the very first time in pharmacy in Pakistan, a first step in promoting and creating gender equity in the profession.

Professional codes of conduct and core competency frameworks (CCFs) within a profession are a means through which EDI can be promoted. These resources are used by schools of pharmacy to develop curricula for pharmacy courses and guide pharmacists in their pursuit of continuing professional development to determine how their practice and learning needs align with standards for the profession. They are also used to assess the competency of pharmacy students through the course of undertaking the pharmacy program. It is therefore important that CCFs developed by pharmacy regulatory bodies around the globe contain statements about the importance of EDI and practicing in a culturally competent manner.

The fact that societies are becoming increasingly multi-ethnic, multicultural, and multi-religious is good. Diversity is a strength, not a weakness.

—Antonio Guterres

Cultural competence training in healthcare education is also needed to ensure that pharmacy professionals can provide patient-centered care to diverse populations. Learning these skills in the classroom and during advanced practice experiences prepares learners to deliver better care and bring more visibility to the profession of pharmacy (Haack, 2008). An example of such training includes the multiple cultural competence eLearning modules for healthcare professionals, with a focus on pharmacists, developed as part of the Inclusive Pharmacy Practice Plan developed by the National Health Service (NHS). An example of one such eLearning module includes "Ramadan: Diabetes and Fasting Toolkit" (National Health Service, 2023b). Such modules allow pharmacists to better understand the healthcare needs of the communities they serve and provide this care accordingly.

The development of resources for pharmacists to better understand and interpret local population health data on health inequalities and better engage diverse communities to design culturally competent and tailored approaches to healthcare delivery, e.g., targeted health promotion, tackling health inequalities through disease prevention, and the management of long-term conditions more prevalent in Black, Asian, and minority ethnic communities, e.g., cardiovascular disease and diabetes, are also of importance. Additionally, when discussing pharmacy professionals, it is important to note that there are often groups within the population that have been economically and socially marginalized due to the lack of cultural competence in education and the workplace. These groups experience adverse impacts on their mental health, work attitude, and job satisfaction, ultimately leading to a reduction in employee retention rates and an overall lack of organizational competitiveness in the marketplace (Scott & Ogden, 2022). Therefore, cultural competence is important for pharmacy professionals to possess not just for the communities they serve but also for their colleagues within the profession.

An RPS initiative titled "Improving Inclusion and Diversity Across Our Profession: Our Strategy for Pharmacy 2020-2025" discusses at length strategies to prioritize and further develop EDI in the UK pharmacy workforce. This strategy was developed because of a survey conducted across the pharmacy profession in the UK. Overall, 66% of respondents agreed that they perceived pharmacy to be a welcoming profession (Royal Pharmaceutical Society, 2020). Examples participants provided of perceived lack of inclusivity in the pharmacy profession included inflexibility of working hours, disability support, and lack of representation in senior positions. The following strategic priorities were developed from the findings of this survey:

- Creating a culture of belonging
- Champion inclusive and authentic leadership
- Challenge inclusion and diversity barriers

Inclusion and diversity work conducted by the RPS focuses on race, gender equality, age, disability, parents and caregivers, Lesbian, Gay, Bisexual, Transgender, Queer, Intersex and Asexual (LGBTQIA+), microaggressions, etc. In February 2023, the Pharmaceutical Society of Australia published its Equality Statement, stating the need for pharmacists to be more inclusive of LGBTQIA+ patients. In the statement, there were five key areas of focus: equitable access to healthcare, education and standards for pharmacists, system changes to facilitate safe and inclusive healthcare, respectful and nondiscriminatory care provided by pharmacists, and the use of inclusive language (Pharmaceutical Society of Australia, n.d.).

Despite the persistent challenges, the past 5 years have witnessed significant improvements in pharmacy related to EDI within healthcare. There continues to be an increasing awareness and commitment from organizations, healthcare systems, and colleges of pharmacy (Sheehan & DiDomenico, 2023). Healthcare organizations have recognized the importance of EDI and have made concerted efforts to address disparities. Many institutions have established EDI committees, implemented training programs, and developed inclusive policies centric to pharmacy personnel. This creates a strategic path forward across all healthcare settings. Collaboration between healthcare organizations, academic institutions, and community stakeholders has strengthened, leading to innovative approaches to address EDI. Strategic partnerships have facilitated outreach efforts and expanded mentorship programs for underrepresented populations (Hovey et al., 2022; White et al., 2022).

Future state, the profession of pharmacy will need to continue to focus on existing challenges persistent to EDI strategies. Despite efforts, there are racial and ethnic disparities in medication access, quality measures, and outcomes. Systemic barriers, including implicit bias and discrimination, contribute to these inequities. As discussed earlier, leadership positions within healthcare still lack adequate representation of minorities and women. Structural barriers and limited opportunities for advancement hinder progress in achieving diverse leadership teams. Pay inequities based on race, ethnicity, and gender also need to be eliminated for EDI to flourish in pharmacy. These disparities contribute to financial inequalities and hinder progress in achieving equitable opportunities for all licensed professionals (Osae et al., 2022).

Experts in the discipline anticipate several trends and forthcoming statistics related to EDI in the future. Pharmacy leaders in healthcare organizations will focus on implementing targeted strategies to address disparities and improve EDI. This includes enhanced recruitment efforts, mentorship programs, and diversity-focused development initiatives. There will be an increasing emphasis on collecting and analyzing data related to EDI outcomes to inform evidence-based interventions. Robust measurement and tracking of EDI initiatives will enable organizations to assess progress and identify areas for improvement. Greater community engagement and partnerships will be established to ensure that EDI initiatives align with the needs and values of the communities being served. By leveraging research, promoting inclusive policies, fostering diverse leadership, and addressing systemic barriers, the profession of pharmacy can create a future where every individual has equal opportunities for success and positive impacts on the lives of all patients.

Equity, diversity, and inclusion: Rationale for systemic action

Increased awareness of the importance of EDI principles has led to efforts to develop these across the industry. In doing so, the aim is toward minimizing bias and discrimination in the profession and the outlook toward society served by the profession. Instilling a framework to promote health equity starts with foundational awareness throughout the continuum of pharmacy practice, from learner to practitioner. Understanding the broader industry impact of positive effects of a strong EDI investment can correlate with organizations with more adaptability and resilience to change, leading to better financial performance, stronger culture and leadership, and more engaged and inspired employees. Efforts undertaken may assist in engaging stakeholders in efforts to improve and further develop leadership commitment, build the organization's employer brand and reputation in the industry, and catalyze progress toward creating a diverse, equitable, inclusive culture for the pharmacy workforce while partnering to drive excellence in safety, quality, experience, and financial outcomes (HBR, n.d.).

Industry trends in EDI efforts

Pharmacy education

Association trends surrounding EDI efforts have increased regarding strategic priorities such as increasing the diversity of applicant pipeline, conducting climate surveys of schools of pharmacy, human capital diversification, and equity in faculty ranks.

Dimensional frameworks for self-ranking, such as the 12 dimensions of the Meyer Diversity, Equity, Inclusion Spectrum tool can help gauge the progress of educational institutions and beyond (Arif et al., 2023). A recent commentary in the *American Journal of Pharmacy Education* highlighted the need for educational outcomes to support overall goals of minimizing health disparities and promoting equity in the communities the profession serves. Efforts at various levels in pharmacy education and accreditation are paramount to drive change forward; implementing an organizational approach offers systemic benefits (White et al., 2022). Standards relating to pharmacy educational outcomes in the Doctor of Pharmacy curricula and postgraduate training residency and fellowship programs should be inclusive of EDI principles. In creating a diverse and socially competent workforce, the opportunity to minimize detrimental health disparities and maximize engagement, innovation, and growth of culture becomes a reality.

Pharmaceutical industry

Pharmaceutical industry-wide efforts to determine where opportunity exists were recently highlighted in the Pharmaceutical Research and Manufacturers of America's (PhRMA) report on The Biopharmaceutical Industry: Improving Diversity & Inclusion in the Workforce. Increased efforts in the pharmaceutical industry through environmental, social, and governance (ESG) efforts have particularly evolved with time. Notable strategies involve efforts to support growth for underrepresented minorities in leadership positions, addressing gender pay gaps, among others (Adams, 2022; The biopharmaceutical industry: Improving diversity & inclusion in the workforce, n.d.). Another key area of focus for the pharmaceutical industry should be increasing diversity in clinical trials. It has long been understood that racial and gender disparity exists in trial recruitment and that this calls into question the generalizability of study findings (Ding et al., 2021). Data show that while gender representation is increasing, work remains to be done on racial representation.

Health system pharmacy practice

Driven by a professional association commitment to increasing awareness through education and engagement with stakeholders, pharmacy divisions and departments are leveraging organizational structure to increase departmental engagement in EDI initiatives. In Europe, the European Association for Hospital Pharmacy has developed a common training framework (CTF). This sets out a set of minimum knowledge, skills, and competencies necessary for the pursuit of the pharmacy profession in the European Union (European Association of Hospital Pharmacists, 2023). The CTF aims to further allow for the facilitation of exchange of expertise, standardization in education quality, and an increase of mobility opportunities for hospital pharmacists across the EU.

Measuring progress and enacting change

Organizational champions, organizational priority, depth of mission, commitment to continuous learning, commitment to diversity and inclusion, and organizational resources, and their key processes for success include change management, information exchange, action research, relationship building, values in action, and leveraging resources (Adams, 2022).

Dimensional frameworks for self-ranking, such as the 12 dimensions of the Meyer Diversity, Equity, and Inclusion Spectrum tool, can help gauge the progress of educational institutions and beyond. The highlighted need for educational outcomes to support the overall goals of minimizing health disparities and promoting equity in the communities the profession serves is prominent. Efforts at various levels in pharmacy education and accreditation are paramount to drive change forward. Standards relating to pharmacy educational outcomes in the Doctor of Pharmacy curricula and postgraduate training residency and fellowship programs can be opportunities to incorporate strategies to include EDI.

A framework from Press Ganey suggests the following strategy and measurement approach (Press Ganey, 2020):

- Align EDI to the mission, vision, and values of the organization; develop a narrative to support EDI; and make direct connections to the organization's or firm's strategic plan.
- Expand recruitment efforts for board members, leadership positions, and caregivers to achieve greater diversity.
- Conduct training for leaders, employees, and medical staff on EDI topics such as recognizing implicit bias, reducing microaggressions, and nurturing cultural competency.
- Ensure that recruitment, performance management, leadership assessment, and training strategies support a more diverse and inclusive organization.
- Collect robust demographic patient and employee data on race, ethnicity, and other key characteristics, and segment key outcome data.
- Measure perceptions of diversity and inclusion among patients and employees and segment the findings by demographics to understand where to focus.

It is summarized that the rationale and strategies of systemic action are based on fundamental goals of minimizing bias, discrimination, and outlook within the profession. Highlighted industry examples demonstrate structural investments in building outcomes that can propel the profession in a direction that promotes positive effects of catalyzing change, higher performance, and strengthened culture and leadership. Engagement assessments can help organizations better assess and benchmark when initial steps are undertaken. Highlights of methods used in pharmacy education and various industries in pharmacy practice to promote awareness, assessment of progress, and identification of needs are shown in Table 1. (See Figs. 1 and 2.)

TABLE 1 Methods and efforts to promote diversity, equity, and inclusion assessment in pharmacy settings.

Application setting	Example of effort/method
Pharmacy education	Outcome(s) development
	Enrollment and applicant statistics
	Climate surveys, assessments—faculty and students
	Accreditation standards for postgraduate residency training and fellowship programs
Pharmaceutical industry	Increase attention on diversity of senior leadership in the firm
	Clinical trial diversity
	EDI leadership (development of boards)
	Addition of employee resource groups
	Improving access and quality of care to medically disadvantaged populations (community benefits, grant programs)
Health system pharmacy	Employee resource groups
	EDI leadership teams
	Community benefits
	Social determinants of health efforts
	Workforce training and education programs
	EDI principles in recruitment of residents and staff

FIG. 1 Alignment for cultural competency.

FIG. 2   Contributing factors to health disparities.

To conclude, the following section lists resources that have been used for supporting cultural competence in the care provided by pharmacists:

Diversity Resource Guide (DRG) for Diversity in Residency Training and the Pharmacy Workforce:
    https://www.ashp.org/-/media/assets/professional-development/residencies/docs/examples/diversity-resource-guide.pdf
Action in Belonging, Culture and Diversity (ABCD):
    https://www.rpharms.com/recognition/inclusion-diversity/abcd
Affirming Pharmacist Care: Understanding Disparities and Creating an Inclusive Environment for Sexual and Gender Minorities:
    https://elearning.ashp.org/products/9728/affirming-pharmacist-care-understanding-disparities-and-creating-an-inclusive-environment-for-sexual-and-gender-minorities
10 Actions for Creating an Inclusive Pharmacy Space:
    https://www.pharmacists.ca/cpha-ca/function/utilities/pdf-server.cfm?lang=en&thefile=/education-practice-resources/SmashingStigma_10Actions.pdf
Incorporating Diversity, Equity and Inclusion into Pharmacy Student, Resident, and Preceptor Training:
    https://elearning.ashp.org/products/9498/incorporating-diversity-equity-and-inclusion-into-pharmacy-student-resident-and-preceptor-training
American Society of Health-System Pharmacists (ASHP) Task Force on Racial Diversity, Equity, and Inclusion: Provides resources on governance, education/training, research, and advocacy in Pharmacy.
    https://www.ashp.org/-/media/assets/about-ashp/docs/DEI-Task-Force-Recommendations.ashx
Harvard Implicit Associations Tests (IAT)—Measures attitudes and beliefs that people may be unwilling and unaware of to report.
    https://implicit.harvard.edu/implicit/takeatest.html

Disclosures

At this time, none of the chapter authors have disclosed any financial conflicts of interest.

References

Adams, B. (2022). *The top 10 Big Pharmas for their diversity, equity and inclusion efforts in 2022.* Fierce Pharma. Retrieved from https://www.fiercepharma.com/special-reports/top-10-equality-diversity-and-inclusion-pharmas-2021.

Arif, S., Butler, L., Gettig, J., et al. (2023). Taking action towards equity, diversity, and inclusion in the pharmacy curriculum and continuing professional development. *American Journal of Pharmaceutical Education, 87*(2), ajpe8902. https://doi.org/10.5688/ajpe8902.

Bukhari, N., Manzoor, M., Rasheed, H., et al. (2020). A step towards gender equity to strengthen the pharmaceutical workforce during COVID-19. *Journal of Pharmaceutical Policy and Practice, 13,* 15. https://doi.org/10.1186/s40545-020-00215-5.

Crews, D., Collins, C., & Cooper, L. (2022). Distinguishing workforce diversity from health equity efforts in medicine. *JAMA Health Forum, 2,* 12. https://doi.org/10.1001/jamahealthforum.2021.4820.

Data USA. (2019). *Pharmacists and pharmacy technicians.* Retrieved from https://datausa.io/profile/soc/pharmacy-technicians#:~:text=The%20workforce%20of%20Pharmacy%20technicians,for%20Pharmacy%20technicians%20is%20White (Accessed 11 August 2023).

Ding, J., Zhou, Y., Khan, M. S., Sy, R. N., & Khosa, F. (2021). Representation of sex, race, and ethnicity in pivotal clinical trials for Dermatological Drugs. *International Journal of Women's Dermatology, 7*(4), 428–434. https://doi.org/10.1016/j.ijwd.2021.02.007.

European Association of Hospital Pharmacists. (2023). *Common training framework—HP education in Europe.* EAHP. Retrieved from https://www.eahp.eu/hp-practice/hospital-pharmacy/common-training-framework.

George, A., Teelucksingh, K., & Fortune, K. (2023). Diversity, equity, and inclusion in the profession of pharmacy: The perspective of three pharmacy leaders. *HCA Healthcare Journal of Medicine, 4*(2), 2. https://doi.org/10.36518/2689-0216.1561.

Haack, S. (2008). Engaging pharmacy students with diverse patient populations to improve cultural competence. *American Journal of Pharmaceutical Education, 72*(5), 124. https://doi.org/10.5688/aj7205124.

HBR. *How investing in DEI helps companies become more adaptable.* Retrieved from https://hbr.org/2023/05/how-investing-in-dei-helps-companies-become-more-adaptable.

Hovey, S., Arif, S., Khan, A., et al. (2022). More than a buzz word: Building diversity, equity, and inclusion into pharmacy residency recruitment. *The Journal of the American College of Clinical Pharmacy, 6,* 2. https://doi.org/10.1002/jac5.1716.

International Pharmaceutical Federation. (2019). *FIP-EquityRx programme.* FIP. Retrieved from https://equityrx.fip.org.

Martin, A., Nauton, M., & Peterson, G. (2021). Gender balance in pharmacy leadership; Are we making progress? *Research in Social and Administrative Pharmacy, 17,* 4. https://doi.org/10.1016/j.sapharm.2020.05.031.

National Health Service. (2023a). *Pharmacy workforce race equality standard report [Internet].* NHS. Retrieved from https://www.england.nhs.uk/long-read/pharmacy-workforce-race-equality-standard-report/#:~:text=The%20publication%20of%20a%20Pharmacy,Pharmacy%20Practice%20(IPP)%20initiative.

National Health Service. (2023b). *Inclusive pharmacy practice.* NHS. Retrieved from https://www.england.nhs.uk/primary-care/pharmacy/inclusive-pharmacy-practice/.

Osae, S., Chastain, D., & Young, H. (2022). Pharmacist role in addressing health disparities—Part 2: Strategies to move towards health equity. *The Journal of the American College of Clinical Pharmacy, 5,* 5. https://doi.org/10.1002/jac5.1594.

Pharmaceutical Society of Australia. *Equality position statement*. Retrieved from https://my.psa. org.au/s/article/Equality-Position-Statement.

Pharmaceutical Society of Ireland. (2023a). *Pharmacy workforce analysis report*. PSI. Retrieved from https://www.thepsi.ie/Libraries/Publications/Pharmacy_Workforce_Survey_Analysis_ Report%C2%AD_September_2023.sflb.ashx.

Pharmaceutical Society of Ireland. (2023b). *Third country route [internet]*. PSI. Retrieved from https://www.thepsi.ie/gns/Registration/Pharmacists/First_Time_Registration_/Non_EU_ Trained_Pharmacists_.aspx.

Press Ganey. (2020). *Diversity and inclusion: Building workforce engagement and improving outcomes in health care [internet]*. Press Ganey. Retrieved from https://www.pressganey.com/wp-content/uploads/2023/02/Building-Workforce-Engagement.pdf.

Royal Pharmaceutical Society. (2020). *Improving inclusion and diversity across our profession: Our strategy for pharmacy 2020–2025*. RPS. Retrieved from https://www.rpharms.com/ LinkClick.aspx?fileticket=plJ2cLMj2H4%3D&portalid=0#:~:text=We%20must%20create %20a%20culture,to%20I%26D%20throughout%20our%20activities.&text=Pharmacists% 20work%20to%20improve%20everyone%27s%20health.

Royal Pharmaceutical Society. (2023). *Inclusion & diversity: Race*. RPS. Retrieved from https:// www.rpharms.com/recognition/inclusion-diversity/race.

Scott, A., & Ogden, R. (2022). Leading diversity, equity, and inclusion efforts within the pharmacy department. *American Journal of Health-System Pharmacy, 79*, 21. https://doi.org/10.1093/ ajhp/zxac215.

Sheehan, A., & DiDomenico, R. (2023). Postgraduate pharmacy training: Much progress and much more to do. *The Journal of the American College of Clinical Pharmacy, 6*, 2. https://doi.org/ 10.1002/jac5.1751.

Smith, J., White, C., Roufeil, L., et al. (2021). A national study into the rural and remote pharmacist workforce. *Rural and Remote Health, 13*, 2. https://doi.org/10.3316/informit.608739189631846.

Tanni, K., & Qian, J. (2021). Trends in gender and race/ethnicity of PharmD students and faculty in US pharmacy schools. *Currents in Pharmacy Teaching & Learning, 13*, 11. https://doi.org/ 10.1016/j.cptl.2021.09.010.

The biopharmaceutical industry: Improving diversity & inclusion in the workforce. Retrieved from https://phrma.org/-/media/Project/PhRMA/PhRMA-Org/PhRMA-Org/PDF/S-U/ TEConomyPhRMA-DI-ReportFinal.pdf.

White, C., et al. (2022). Organizational commitment to diversity, equity, and inclusion: A strategic path forward. *American Journal of Health-System Pharmacy, 79*(5), 351–358. https://doi.org/ 10.1093/ajhp/zxab322.

Chapter 9

Moving in complex spaces: A call to action for equity, diversity, and inclusion in physiotherapy

Stephanie N. Lurch[a,b,c,d] and **Jeffrey John Andrion**[a,e,f,g]

[a]*Department of Physical Therapy, University of Toronto, Toronto, ON, Canada,* [b]*School of Rehabilitation Science, Physiotherapy Program, McMaster University, Hamilton, ON, Canada,* [c]*York Catholic District School Board, Aurora, ON, Canada,* [d]*Children's Treatment Network, Richmond Hill, ON, Canada,* [e]*Department of Physical Therapy, A.T. Still University, Mesa, AZ, United States,* [f]*A.T. Still Research Institute, A.T. Still University, Mesa, AZ, United States,* [g]*Philippine Working Group—International Centre for Disability and Rehabilitation, University of Toronto, Toronto, ON, Canada*

Introduction

There is one thing that unites all human beings around the globe: we all want a long, happy, healthy, full, and free life for ourselves and our loved ones. Movement is crucial to this equation. It is crucial for health and well-being. It is crucial for development. Our neurons, our heart, our ideas, our relationships, our professions, our institutions, our systems are in perpetual motion. It's time we started thinking about movement as something more than just bodily movement.

Lurch (2021)

As movement experts, the above epigraph reflects the depth and breadth of what it means to move. It both challenges and provides an opportunity to re-imagine how movement can be interpreted. This chapter on equity, diversity, and inclusion (EDI) in physiotherapy (PT) is a call to action—that is, an imperative to move. It serves to interrogate the PT profession from an EDI perspective, provide opportunities to think critically about our past and present, and offer insights to help propel us toward a more equitable, diverse, and inclusive future. We foreground systemic racism in this discussion. We call attention to the systemic marginalization of racialized people throughout the organizations and systems that make up the PT profession. We aim to ignite actions that drive positive changes in physiotherapy and the world at large.

Equity, Diversity, and Inclusion in Healthcare. https://doi.org/10.1016/B978-0-443-13251-3.00009-0

The origin of physical therapies can be traced back 5000 years to Northern Africa, where the excavation of ancient Egyptian burial sites uncovered remains with fractures that had been bound with splints of bark, wrapped in linen, and held in place by bandages (Ellis, 2011; Marshall, 2022). Similarly, as early as 3000 BCE, there is evidence that suggests the Chinese engaged in massage as a way of healing (Moffat, 2012). This challenges the notion that physiotherapy only started a century ago.

Physiotherapy is a globally recognized, evidence-informed health profession within the discipline of rehabilitation sciences. In 2020, there were 1,901,585 physiotherapists practicing worldwide, 63% of whom were female (World Physiotherapy, 2020). From managing various conditions in primary care settings (Cott et al., 2011; Vader et al., 2022) to responding to global health crises (Antony Leo Asser & Soundararajan, 2021; Nixon et al., 2010), physiotherapists are recognized as important healthcare team members. In Canada, PT is a regulated health profession, born out of a need to rehabilitate injured soldiers returning from World War I (Evans, 2010).

The entry-to-practice educational requirements for PTs graduating in Canada have evolved from a diploma to a master's degree or equivalent (The Canadian Council of Physiotherapy University Programs/Le Conseil canadien des programmes universitaires de physiothérapie, 2023). In the United States (US), the minimum entry-level requirement is a clinical doctorate degree. Despite its evolution, the profession has significant challenges that have sociohistorical roots.

Whiteness, white supremacy, and racism

A growing body of scholarship on EDI in physiotherapy points to a profession that is oriented toward whiteness (Hammond et al., 2019; Hughes et al., 2021; Matthews et al., 2021; Vazir et al., 2019; Wegrzyn et al., 2021). The term whiteness refers to "practices, policies, and perspectives that enable the dominance of white people, norms and culture, in institutions, systems and society" (National Collaborating Centre for Determinants of Health, 2020). White supremacy is the institutionalization of whiteness and white privilege (York University, 2022). Similar to Arcobelli (2021), in this chapter, we refer to white supremacy as a system versus an individual conviction or belief. That is, it refers to the political, economic, and cultural systems in which whites control power, resources, and ideas (York University, 2022).

White is a color (Caivano, 2022) and a race. Race is a social construct for which there is no biological basis (Delgado & Stefancic, 2017). As a construct, it can be invented, changed, and abandoned (Steinberg, 2005). The white race is seen as neutral and is a designation against which all other races are compared. White supremacy, the belief that white people are superior and people of other races are inferior, is at the root of systemic racism. Systemic racism is broadly understood as policies and practices that unfairly advantage some and disadvantage others based on race (Cambridge Dictionary, n.d.). Systemic racism is entrenched in society, reproduced, and reinforced. It has been a successful way of pathologizing difference and is "a pernicious and vicious social problem" (Dei, 2005, p. 138). We will demonstrate that it is a problem in physiotherapy.

On movement

The phenomenon of movement is based on an object occupying different positions in space at various times. A theory of movement therefore includes a theory of position, time, objects, and space (Galton, 2007) that can influence each other, overlap, and be iterative. The concepts of space and time include ideas from philosophy and the social sciences, where the boundaries of space and time merge, moving us from the absolute to relative to relational spaces (Harvey, 2006). These remind us that our past is our present, and these together inform our future. We metaphorically use this theory of movement as a conceptual framework to organize our thinking. We focus on the theory of space, informed by distinguished scholar and critical geographer Harvey (2006), as the central analytical framework in this chapter.

As we aim to examine the role of power and unearned structural advantage in PT, our work uses the critical theory framework to examine the PT profession. The critical theory aims to reveal, critique, and challenge the powerful structures that cause inequities (Lincoln et al., 2011). Most importantly, we strive to change these structures in keeping with the activist dimension of critical theory. As we begin using these three frameworks, critical reflection will be the lens that informs and shapes our analysis, in alignment with PT training and clinical practice. Critical reflection is a method to examine taken-for-granted ideas and concepts (Gibson et al., 2010). The theory of space has been a neglected analytical framework within the health professions (Gregory et al., 2014; Van Vuuren & Westerhof, 2015) that we think is of value to the examination of EDI issues in PT.

Position: Social locations

A position refers to where someone or something is in space. It is important for us at the outset to present who we are as authors and how our individual and collective perspectives—our social locations—shape our understanding of EDI issues in PT. We bring diverse perspectives, experiences, and identities to this work. We acknowledge that we have both relative privilege and disadvantage in this space. Like our profession, our social locations are in flux, and we are careful to avoid the binary thinking that dangerously simplifies our positions and acts as a barrier to more nuanced understandings. We draw on bell hooks'[a] concept of seeing the world while centrally located and at the same time from the margin (Yoshida et al., 2016). hooks contends that "to be in the margin is to be part of the whole but outside the main body" (Hooks, 1984, p. ix). This

---

a. Bell Hooks does not capitalize her name in order to shift the focus from herself to her writing.

provides a unique perspective from which to see the world from "both the outside in and the inside out" (Hooks, 1984, p. ix). These are our positions as authors. As physiotherapists, we are part of the main body of PT but also belong to intersecting equity-denied communities, which place us squarely in the margin. Though we are Canadian, our people are indigenous to Central and West Africa, Europe, the Philippines, and the Caribbean. We are scholars, university faculty, clinicians, learners, community members, activists, and antioppression educators. We each have our own histories, languages, memories, and identities. We are Black and brown, male and female, able-bodied, cis-gendered, heterosexual, Christian, spiritual, and from humble beginnings. We are not all-knowing. Neither are we unbiased. What we have in common is this: a commitment to truth, justice, and humanity. Ultimately, our interest in using movement phenomena with a particular focus on space theory has been inspired by our innate curiosity to fully understand how our profession's history, position, and trajectory are impacted by past and contemporary hegemonic social forces. We believe that EDI work is about people, and consequently, we invite our readers to also imagine how this "space" will help illuminate and spark change within our profession.

Time and position: The historical nature of power

Direction and duration are the fundamental characteristics of time. Scholar and sociologist Quijano (2000) states:

> *The future is an open temporal territory. Time can be new, and so not merely the extension of the past. And in this way history can be perceived now not only as something that happens… but also as something that can be produced by the action of people…(p. 547).*

Systems of oppression necessitate an analysis that is both historical and contextual (Dei, 2006). Importantly, these systems also warrant an analysis of power as illustrated by Quijano (2000) in his conceptual framing of the colonial matrix of power and Western modernity which created a new temporal perspective of history. This framing demonstrates how hierarchical structures—racial, political, social, and economic—shape the modern world (Mignolo, 2011; Quijano, 2000).

Race, racialization, and racism are embedded in institutions, in society, in healthcare, and in PT (Hughes et al., 2021; Vazir et al., 2019; Wegrzyn et al., 2021). Race did not have a meaning before the colonization of America (Quijano, 2000). With the advent of colonization, Quijano argues that two historical processes informed a new model of global power. The first was the classification of the world's population around race, which positioned white Europeans as biologically superior and racialized non-Europeans as inferior. Race became the standard that separated people into ranks and roles—and has proven to be the longest-standing instrument of social domination, eclipsing gender domination. The second was the control of labor, resources, and products. Europeans associated unpaid labor with inferior races/non-Europeans and paid labor with superior races/Europeans. Over time, a systematic racial distribution of labor was implemented, which impacted world capital markets. This has impacted healthcare delivery. For instance, the evolution of global health demonstrates how racialized bodies were used to implement the various colonial projects of Westerners in Africa and Asia (Greene et al., 2013). Further, the new model of global power concentrated the control of knowledge and knowledge production in the West (Paton et al., 2020; Quijano, 2000). Grosfoguel (2013) argues that the seed of epistemic power rests with white males from France, England, Germany, Italy, and the United States. This positioned European knowledge as superior/modern and non-European knowledge as inferior/primitive. Importantly, the colonized had their own ways of knowing, meaning making, symbolism, and objective and subjective realities replaced by those of the colonizers (Maldonado-Torres, 2007; Ndlovu, 2018). Hence, colonization occurred both through seizing and displacing peoples from their land and through the colonization of minds (Asante, 2006). This has negated the value of forms of knowing, understanding, seeing, and doing held by the colonized (Ndlovu, 2018).

Object: The physiotherapy profession

According to movement theory, anything must count as an object (Galton, 2007). Our object or unit of analysis is the PT profession in its entirety. The most prominent international body for the profession, World Physiotherapy (n.d.), defines diversity as "having people with different characteristics in a given setting." Hammond et al. (2019) described the default PT student identity as white, middle class, mature, and female. Notably, in Canada, evidence is emerging as to the critical underrepresentation of PT students, and subsequently practitioners, from historically marginalized groups, particularly those from racialized communities (Hughes et al., 2021; Vazir et al., 2019; Wegrzyn et al., 2021). The findings in these recent Canadian studies are supported by similar studies in the United Kingdom (UK) (Yeowell, 2013), the US (Matthews et al., 2021), and South Africa (Cobbing, 2021). Physiotherapy spaces are characterized by whiteness and the assumption of white superiority (Arcobelli, 2021). Wegrzyn et al. (2021) further report that these spaces are also characterized by anti-Blackness, with Black students experiencing exclusion, discrimination, and low academic expectations from faculty.

Within the North American PT education system, not only is diversity limited, but so too are equity and inclusion (Hughes et al., 2021; Matthews et al., 2021). For example, a Canadian study by Cox et al. (2019) identified systemic barriers in the recruitment of potential PT students from Indigenous communities. Racialized students in various Canadian PT programs have revealed instances of internalized, personally mediated, and institutionalized racism (Hughes et al., 2021), as described by Jones' three-level framework of racism

(Jones, 2000). Racialized PTs are disadvantaged in many areas, including leadership, education, and employment (Vazir et al., 2019). For instance, Andrion (2022) reports that racialized transnational physiotherapists (TNPs)[b] feel ghettoized to work in practice areas and settings that are normally shunned by Canadian-educated colleagues. Moreover, while there is no definitive demographic data regarding the leadership of academic programs and regulatory colleges in Canada, anecdotal evidence indicates racialized minorities are not leading classes and are missing in the boardroom, in the classroom, in the workplace, and in the profession at large. Similar findings are noted in the US (Matthews et al., 2021).

Paton et al. (2020) explain that the foundation of health profession curricula is built on systems of oppression that include racism, sexism, patriarchy, authority, and capitalism and is grounded in Western, Eurocentric epistemology. In this way, healthcare education is clearly not a time-limited event but rather a structure. There are concerns that coloniality continues to repeat itself through health professions education. Over the course of decades, professionals and critical scholars have called for decoloniality, including scholars whose orientation is toward a reflexive approach to healthcare and healthcare education, specifically PT education (Arcobelli, 2021; Blake, 2020; Cobbing, 2021; Gasparelli, 2019; Gibson et al., 2010; Nixon, 2019). For example, Cobbing (2021) writes extensively on the Eurocentric homogeneity of PT curricula. In Canada, we are hopeful that curriculum changes to meet new accreditation standards will gain momentum. Notably, a demonstration of relational accountability to Indigenous peoples and an orientation toward justice-driven and anti-oppressive practices are now required to comply with the entry-to-practice physiotherapy education program accreditation standards (Physiotherapy Education Accreditation Canada, 2020).

From a clinical practice standpoint, it appears that patients from the rural, remote, and northern areas of Canada have little or no access to PT services (Shah et al., 2019). Indeed, like the default PT identity, there is a default PT service user identity influenced by the determinants of health. A systematic review by Braaten et al. (2021) revealed those who use physiotherapy services tend to be female, employed, live near an urban center, and have insurance, higher levels of education, and a higher income. In the U S, non-Hispanic whites are more likely to seek PT. Those who face barriers to service have lower levels of education, live in rural areas, and are of low socioeconomic status. We also contend that not all patients are treated the same. Anti-Black and anti-Indigenous racism negatively affects the quality-of-care members of these communities receive (Douglas et al., 2022; Turpel-Lafond, 2020).

---

b. Coined by Andrion (2022), we feel that this term is more encompassing versus the commonly used ones, such as internationally educated or foreign-trained.

Space: Three dimensions

We intend to explore how EDI and *othering* in the PT profession have been influenced by temporospatial events, to analyze the sociohistorical context of PT and identify how that context contributes to the EDI challenges in our profession. The work on spatial theory refers to "objects" as units of analysis. As explained earlier, we identify the profession as the object in our analysis, like Naidu's (2021) interpretation of modern medicine as a colonial artifact.

Harvey (2006) differentiates space into three levels: absolute, relative, and relational (Fig. 1). **Absolute** refers to space that is fixed, asocial, bounded by territories and time, and where things remain similar and immovable. **Relative space** is the relative position of one object in comparison to another and "depends crucially upon what it is that is being relativized and by whom" (p. 272). It is not independent of time. Finally, **relational space** is contained within objects and exists through the combination of social and spatial relations between objects. Relational space is socially made. It is reshaped over time (Warf, 2010). It is influenced by internal and external factors and depends on everything else occurring around it (Harvey, 2006).

Absolute space

As we examine EDI in PT, we must first understand how the profession was recognized officially and by whom. The absolute PT space exists because physiotherapy exists. As we reflect on this space, we consider how history and time are immovable elements. Even in its infancy, the PT profession, as we know it, with its own set of standards, rules, legislation, and proclamations, established its own boundaries within the medical profession. The creation of the physiotherapy "territory" allowed it to set itself somewhat apart from other health

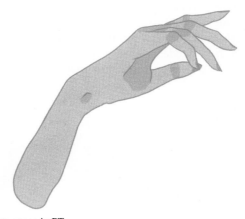

FIG. 1   The absolute space in PT.

professions. From this standpoint, the PT absolute space was and is distinctively white (Evans, 2010; Hammond et al., 2019; Matthews et al., 2021; Moffat, 2012), as depicted in Fig. 1.

Over the course of history, white spaces have become spaces of power and influence (Delgado & Stefancic, 2017; Hooks, 1984; Paton et al., 2020; Quijano, 2000). As our exploration of PT's history reveals, physiotherapy has not been spared. The absolute space—space that is fixed, bounded by territories, and where things remain similar and immovable—of the profession is a white space by design, and this white absolute space holds power and influence.

Relative space

Relative space is characterized by the intimate connection between time and space and the positioning of objects relative to each other (Harvey, 2006). This space is organized by forces of competition such that the relative positioning of an object is influenced by power structures (Kesteloot et al., 2016). PT occupies relative space because it exists relative to other health professions.

With its boundaries well established in the absolute white space, PT become an autonomous profession. It used its own position within this space not only to interact relative to the other professions but, more importantly, to take advantage of how it could further expand the territory that was first defined in the absolute space. For example, in addition to the change in entry-to-practice education (from a diploma in the absolute space to a master's or doctorate level in the relative space), we have also seen changes in clinical care, including the ordering of special tests (Keil et al., 2019), direct access to care (Demont et al., 2021), and advanced practice models (World Confederation for Physical Therapy, 2019), previously the purview of physicians only.

In the relative space, as in the absolute space, we see from a sociohistorical standpoint that the PT profession remains the sphere of white, Western practitioners relative to *others*, as depicted in Fig. 2. Notably, the positioning between objects in this space also means unequal distributions of power. For instance, it is well documented that categories of historically grounded intersecting systems of oppression such as race, gender, sexual orientation, age, size, class, faith, and ability shape people's experiences and health through simultaneous rather than

FIG. 2 The relational space in PT.

additive interactions (Crenshaw, 1991). The interconnected nature of these social categorizations creates intersecting systems of disadvantage that make it less likely for people from equity-denied communities to enter and thrive in PT (Hammond et al., 2019; Hughes et al., 2021; Matthews et al., 2021; Vazir et al., 2019; Wegrzyn et al., 2021). While there are no studies documenting racial disparities in PT outcomes for clients (Matthews et al., 2021), we argue racial disparities in health outcomes found in medicine likely apply (Datta et al., 2021).

Relational space

Relational space involves a combination of social and spatial relations (Kesteloot et al., 2016), as depicted in Fig. 3. Given our knowledge of the history and position of the PT profession and our exploration of the absolute and relative spaces, we see that the present and future states of the profession in the relational space are dependent on the influence and impact of social

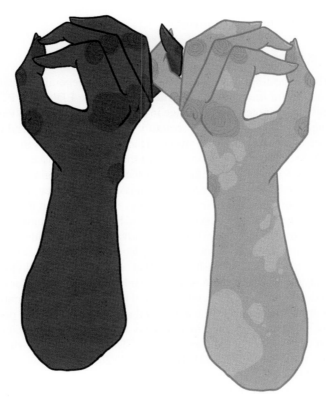

FIG. 3 Relational space in PT.

relationships between people, institutions, systems, and forces acting on them. Harvey (2004) speaks about relational space in this way:

> *An event or a thing at a point in space cannot be understood by appeal to what exists only at that point. It depends upon everything else going on around it (although in practice usually within only a certain range of influence). A wide variety of disparate influences swirling over space in the past, present and future concentrate and congeal at a certain point to define the nature of that point. (p. 4)*

Kesteloot et al. (2016) argue relational space comes into existence when entities consider it their space. Pragmatically, this means the PT profession has the power to shape and fill *our* space as we please. The practice of PT is human work. Human work is relational work. Society is built on relationships that are shaped by dominant structures (i.e., social relations) between individuals, groups, institutions, and systems that make up a community. In other words, hierarchies within the relational space determine who does what, who gets what, who knows what, and who decides. Building on Ndlovu's (2018) insights, we argue that relational space has both been resistant to change and susceptible to rearrangement. We further argue that relational space in PT must be radically transformed rather than just rearranged.

So, where does this take us?

We began by suggesting how movement phenomena in general and spatiality in particular are useful frameworks for critically analyzing EDI in the PT profession. Because of the potential for change in relational spaces, we are tempted to focus our analysis on it, given how spatiality puts emphasis on social relationships, processes, and structures. However, Harvey (2006) makes the important argument that relational space can embrace the other two spaces and that, in reality, the three spaces should be kept in constant dialectical tension where they interplay with each other. Like movement, the concept of space, when merged with time, position, and object, resides in a universe of constant discourse on a continuum between the past, present, and future; between the imagined and real; and between those with unearned structural advantage and disadvantage.

We argue that power and unearned privilege are key features in all three spaces. These have been established, guarded, and granted to individuals and institutions, typically to maintain whiteness and white supremacy for political, cultural, and economic gain (Kendi, 2017). Therefore, our past continues to shape our current state. That is, the norms and standards of the early profession as we know it had an orientation to whiteness at a time when sentiments were high against racialized people (Lacroix, 2016). This orientation to whiteness in PT continues today (Cobbing, 2021; Hughes et al., 2021; Matthews et al., 2021; Vazir et al., 2019; Wegrzyn et al., 2021).

With respect to the relative space, as PT established itself within the medical field, it grew beyond the absolute space and found ways to make itself more relevant compared to other medical disciplines. This was achieved by demanding higher academic requirements of practitioners and growing their areas of expertise and practice by mirroring some of the work of those in the medical profession. As the PT profession expands its areas of expertise, it also grows in size; however, it should be noted how physiotherapists were and remain positioned relative to each other during this growth. Racialized bodies remain excluded. Canada has imported racialized physiotherapists from countries with limited resources; however, they can largely only gain access to the profession if they assimilate and have "Canadian experience" (Andrion, 2022). The absolute space can and does impact the relative space.

As we enter the relational space, we observe three hegemonic forces—neoliberalism, globalization, and the legacy of colonialism—that appear to perpetuate the sociohistorical context upon which the PT profession was built. The connivance of these three elements perpetuates systemic racism in the PT profession. First, neoliberalism, with its principles of deregulation and privatization (Ostry et al., 2016), has influenced directly or indirectly the practice of the physiotherapist, that is, who gets to practice and who gets served (Braaten et al., 2021). Globalization and the privatization of health services have resulted in the migration of racialized physiotherapists and other healthcare professionals (Cornwall et al., 2015) from low-income countries to serve the West. However, when they arrive, racialized transnational physiotherapists face significant barriers to licensure, which limit their prospects of practicing as physiotherapists (Andrion, 2022). This form of racial capitalism (Robinson, 2000) maintains the whiteness of the profession and ensures the steady supply of brown and Black bodies to support Western healthcare systems in more menial jobs (Andrion, 2022). One can also appreciate how these racialized bodies have become commodities (export products) to their home countries. The Philippines and India are two of the top producers of healthcare professionals globally, including PTs (Lorenzo et al., 2005), where their remittances contribute to national economic growth.

As we consider the intergenerational effect of colonialism, its legacy continues. PT education in India and the Philippines are classic examples, patterned after the British and American systems, respectively. More than this, once these Western-educated (colonialism) racialized bodies complete university, they are exported (globalization) to high-resource countries whose privatized healthcare systems (neoliberalism) recruit healthcare professionals from low- to high-resource settings to serve Western patients. Thus, the imported Western-based educational system has become fertile ground for neocolonialism.

Under these circumstances, movement and changes to PT space become necessary. What if these spaces had intentionally punctured borders? Would such perforations create the space needed to initiate the movement of racialized PTs from the margin toward the center? Would porous borders encourage the development of professionals and the profession relative to others? Would relationships at the individual, institutional, and systemic levels thrive? We are convinced that, unless such perforations coupled with targeted actions occur, the object (the PT profession) will not change direction. In keeping with the laws of motion, the object needs to be acted upon by an unbalanced and intentional force. This force is our call to action.

The relational space must respond to the intentional forces of change; otherwise, addressing contemporary EDI issues in PT will be problematic. Why? Racialized bodies are excluded and subjected to systemic oppression, in particular racism, and injustice in the absolute space. Second, because of colonialism, these same bodies are viewed as commodities in the relative space. Third, these racialized physiotherapists continue to be exploited within the relational space.

We believe the PT profession is *already* occupying this relational space and worry about the perpetuation of white supremacy in this space. Will the PT profession accept the challenge to initiate a paradigm shift? Will physiotherapists change the landscape within the relational space? Or will they continue to ignore the sociohistorical context of the profession's development and limit its capacity and borders to the detriment of the profession, its racialized practitioners, and the communities it serves?

As we reflect on the tension between the three spaces, we are further convinced that the relational space is indeed the focal point of contestation. In other words, while there are things that we can no longer change, such as the history of the PT profession, there is an opportunity for physiotherapists to act *now* and address the systemic oppression experienced by their racialized colleagues and patients. The relational space in physiotherapy must be radically transformed rather than rearranged as depicted in Fig. 4, which symbolically reflects freedom, hope, and new beginnings. In order to do this, we must puncture the borders and co-imagine a radically different future where racialized PTs and patients occupy the center, claiming and sharing in the power and privilege that

FIG. 4 A radically transformed relational space in PT.

they rightfully deserve. Once this happens, physiotherapy will become a profession that is true, just, and humane for all and not solely for a few. EDI will then inhabit the three spaces rather than be blocked by rigid borders. It is imperative that we move from spaces of whiteness and power to spaces that embrace and restore communities that have been harmed. This call to action acknowledges that the heavy lifting required should not be borne by members of racialized communities who have suffered and already understand why systemic change needs to happen. White supremacy is destructive to everyone, even those who most benefit from it (National Collaborating Centre for Determinants of Health, 2020). We say to our privileged colleagues, will you act?

Movement: From theory to action

We acknowledge that there is no single way to create a more equitable, diverse, and inclusive PT profession. Hassen et al. (2021) contends there is a paucity of data in terms of what types of antiracism interventions work in healthcare settings (and we add health professions education environments). There is some evidence to suggest that simultaneous interventions at systems, organizational, interpersonal, and individual levels are needed to make sustainable change (Hassen et al., 2021). EDI scholar Dr. Leeno Karumanchery (personal communication, May 23, 2023) argues these interventions must be "systemic, sustainable, and scalable." We believe an approach guided by the action-oriented dimensions of critical theory may be key. We provide four actions for conducting EDI work in PT, informed by scholars and members of equity-denied communities. We ask that you simply practice with CARE:

> **C**hampion change across all three spaces (absolute, relative, and relational spaces) and at all three levels of racism (internalized, personally mediated, and institutional). The complexity of racial issues necessarily requires a multistrategic approach (Hassen et al., 2021). This may mean ceding space to members of equity-denied communities who experience more barriers.
> **A**ction critical approaches to EDI. In keeping with critical theory, this may include embracing critical whiteness as an antiracist practice. Critical whiteness problematizes white as the norm. It illuminates how power operates to perpetuate white racial domination, the role of whiteness in creating inequities, and shifts the focus away from those who are racialized to those who are not (National Collaborating Centre for Determinants of Health, 2020). Educate yourself. Critical education, in particular antiracist education, is a proactive practice (Dei, 2005) that addresses all forms of racism and is a way to analyze, respond, prevent, and dismantle it.
> **R**e-distribute power. Transforming governance, institutional policies, practices, and entrenched norms are systems-level factors that can shift power if divorced from whiteness and white supremacist ideology and actions (National Collaborating Centre for Determinants of Health, 2020). Unless we understand how power is implicated in perpetuating the challenges faced by equity-denied communities, we will not be able to ensure equity for all.

Engage community. Hassen et al. (2021) found that only 21% of peer-reviewed articles about antiracist action included an antiracism intervention at the community level. There is evidence to support the importance of developing meaningful relationships with racialized communities and patient populations at multiple intervention levels to address community-prioritized issues. Authentic relationship building with community is a way forward in accordance with the principle "nothing about us without us." This means developing an allied space. Bishop (1994), Goodman (2015), and Nixon (2019) report on the importance of critical allyship, engaging a network of individuals who may not share the same lived experience but who are genuinely committed to ensuring that social justice is achieved by all.

Conclusion

In this chapter, we hope that we have succeeded in demystifying EDI issues in the PT profession. While power and privilege have shaped PT throughout the past 100 years, they must not shape our future. Our analysis using movement phenomena with a focus on the spaces physiotherapy occupies helps to highlight and understand that systemic oppression, in particular racism, in PT is deeply entrenched. We worry that if neoliberalism, globalization, and coloniality are left unchecked, the oppression experienced in relational space will continue to grow. We are hopeful, however, that PT will begin to perforate these borders to allow those at the margins to move toward the center. Our analysis shows that the PT profession in Canada has come to an age of maturation, whereas as a profession, we need to address these hegemonic forces if our goal is to have a diverse, equitable, and inclusive profession: not just for some but for all.

We've found what moves us. What will move you?

Acknowledgments

The authors are forever grateful to Kathy Davidson, Michelle Lurch-Shaw, and Shaun Cleaver for their wisdom, generosity, transparency, and editorial contributions. The authors thank Jordyn Coles for her poignant illustrations. The authors are indebted to and honor those racialized PTs who came before them and upon whose shoulders they stand.

Disclosures

There are no conflicts of interest.

References

Andrion, J. J. A. (2022). *The adaptation experiences of transnational physiotherapists in Ontario, Canada: A grounded theory approach* (PhD dissertation) https://yorkspace.library.yorku.ca/xmlui/bitstream/handle/10315/39719/Andrion_Jeffrey_JD_PhD_2022.pdf?sequence=2&isAllowed=y.

Antony Leo Asser, P., & Soundararajan, K. (2021). The vital role of physiotherapy during COVID-19: A systematic review. *Work, 70*(3), 687–694. https://doi.org/10.3233/WOR-210450. PMID: 34719461.

Arcobelli, L. (2021). *Physiotherapy curricula and Indigenous peoples: A snapshot of Canadian physiotherapy programs* (thesis) https://escholarship.mcgill.ca/downloads/5h73q1782.

Asante, M. (2006). Foreword. In G. J. Sefa Dei, & A. Kempf (Eds.), *Anti-colonialism and education: The politics of resistance* (pp. ix–x). Rotterdam, Netherlands: Sense Publishers.

Bishop, A. (1994). *Becoming an ally: Breaking the cycle of oppression.* Fernwood Publishing.

Blake, T. (2020, July 9). *The internet is free: Progressing past awareness towards racial justice in physiotherapy [Webinar].* Embodia. https://embodiaapp.com/webinars/185-the-internet-isfree-progressing-past-awareness-and-towards-racial-justice-in-physiotherapy.

Braaten, A. D., Hanebuth, C., McPherson, H., Smallwood, D., Kaplan, S., Basirico, D., Clewley, D., & Rethorn, Z. (2021). Social determinants of health are associated with physical therapy use: A systematic review. *British Journal of Sports Medicine, 55*(22), 1293–1300. https://doi.org/10.1136/bjsports-2020-103475.

Caivano, L. (2022). Black, white, and grays: Are they colors, absence of color or the sum of all colors? *Color Research and Application, 47*(5), 252–270. https://doi.org/10.1002/col.22727.

Cambridge Dictionary. (n.d.). Systemic racism. https://dictionary.cambridge.org/dictionary/english/.

Cobbing, S. (2021). Decoloniality in physiotherapy education, research, and practice in South Africa. *The South African Journal of Physiotherapy, 77*(1), 1–6. https://doi.org/10.4102/sajp.v77i1.1556.

Cornwall, M. W., Keehn, M. T., & Lane, M. (2015). Characteristics of US-licensed foreign-educated physical therapists. *Physical Therapy, 96*(3), 293–304.

Cott, C. A., Mandoda, S., & Landry, M. D. (2011). Models of integrating physical therapists into family health teams in Ontario, Canada: Challenges and opportunities. *Physiotherapy Canada, 63*(3), 265–275.

Cox, J., Kapil, V., McHugh, A., Sam, J., Gasparelli, K., & Nixon, S. A. (2019). Build insight, change thinking, inform action: Considerations for increasing the number of Indigenous students in Canadian physical therapy programmes. *Physiotherapy Canada, 71*(3), 261–269.

Crenshaw, K. (1991). Mapping the margins: Intersectionality, identity politics, and violence against women of color. *Stanford Law Review, 43*(6), 1241. https://doi.org/10.2307/1229039.

Datta, G., Siddiqi, A., & Lofters, A. (2021). Transforming race-based health research in Canada. *Canadian Medical Association Journal, 193*(3), E99–E100. https://doi.org/10.1503/cmaj.201742.

Dei, G. J. S. (2005). Unmasking racism. In L. Karumanchery (Ed.), *Engaging equity: New perspectives on anti-racist education* (pp. 135–148). Detselig Enterprises.

Dei, G. J. S. (2006). Introduction: Mapping the terrain-towards a new politics of resistance. In G. J. S. Dei, & A. Kempf (Eds.), *Anti-colonialism and education: The politics of resistance.* Rotterdam: Sense Publishers.

Delgado, R., & Stefancic, J. (2017). *Critical race theory: An introduction.* New York University Press.

Demont, A., Bourmaud, A., Kechichian, A., & Desmeules, F. (2021). The impact of direct access physiotherapy compared to primary care physician led usual care for patients with musculoskeletal disorders: A systematic review of the literature. *Disability and Rehabilitation, 43*(12), 1637–1648.

Douglas, D., Ndumbe-Eyoh, S., Osei-Tutu, K., Hamilton-Hinch, B.-A., Watson-Creed, G., Nnorom, O., & Dryden, O. H. (2022, October 24). Black Health Education Collaborative:

The important role of Critical Race Theory in disrupting anti-black racism in medical practice and education. *CMAJ*. https://www.cmaj.ca/content/194/41/E1422.

Ellis, H. (2011). The early days of splints and splinting. *Journal of Perioperative Practice, 21*(7), 251–252. https://pubmed.ncbi.nlm.nih.gov/21874991/.

Evans, S. (2010). Coming in the front doors: A history of three Canadian physiotherapists through two World Wars. *Canadian Military History, 19*(2), 55–62.

Galton, A. (2007). Between space and time. In O. Stock (Ed.), *Spatial and temporal reasoning* (pp. 321–352). Netherlands: Springer.

Gasparelli, K. (2019). Clinician's commentary on Oosman et al. *Physiotherapy Canada, 71*(2), 158–159. https://doi.org/10.3138/ptc.2017-94-cc.

Gibson, B. E., Nixon, S. A., & Nicholls, D. (2010). Critical reflections on the physiotherapy profession in Canada. *Physiotherapy Canada, 62*(2), 98–100.

Goodman, D. J. (2015). Oppression and privilege: Two sides of the same coin. *Journal of Intercultural Communication, 18*, 1–15.

Greene, J., Basilico, M. T., Kim, H., & Farmer, P. (2013). Colonial medicine and its legacies. In P. Farmer, J. Y. Kim, A. Kleinman, & M. Basilico (Eds.), *Reimagining global health* (pp. 33–73). University of California Press (Chapter 3).

Gregory, L. R., Hopwood, N., & Boud, D. (2014). Interprofessional learning at work: What spatial theory can tell us about workplace learning in an acute ward. *Journal of Interprofessional Care, 28*(3), 200–205.

Grosfoguel, R. (2013). The structure of knowledge in westernized universities: Epistemic racism/ sexism and the four genocides/epistemicides. *Human Architecture: Journal of the Sociology of Self-Knowledge, XI*(1), 73–90.

Hammond, J. A., Williams, A., Walker, S., & Norris, M. (2019). Working hard to belong: A qualitative study exploring students from Black, Asian and minority ethnic backgrounds experiences of pre-registration physiotherapy education. *BMC Medical Education, 19*(1), 1–11.

Harvey, D. (2004). *Space as a key word David Harvey—Front desk apparatus*. https:// frontdeskapparatus.com/files/harvey2004.pdf.

Harvey, D. (2006). Space as a keyword. In N. Castree, & D. Gregory (Eds.), *David Harvey: A critical reader* (pp. 70–93). Blackwell Publishing (Chapter 14).

Hassen, N., Lofters, A., Michael, S., Mall, A., Pinto, A. D., & Rackal, J. (2021). Implementing anti-racism interventions in healthcare settings: A scoping review. *International Journal of Environmental Research and Public Health*. https://pubmed.ncbi.nlm.nih.gov/33803942/.

Hooks, B. (1984). *Feminist theory: From margin to center*. Boston, US: South End Press.

Hughes, N., Norville, S., Chan, R., Arunthavarajah, R., Armena, D., Hosseinpour, N., Smith, M., & Nixon, S. A. (2021). Exploring how racism structures Canadian physical therapy programs: Counter-stories from racialized students. *Journal of Humanities in Rehabilitation*. https:// www.jhrehab.org/2019/11/14/exploring-How-racism-structures-canadian-physical-therapy-programs-counter-stories-from-racialized-students/.

Jones, C. (2000). Levels of racism: A theoretic framework and a gardener's tale. *American Journal of Public Health, 90*(8), 1212–1215.

Keil, A. P., Baranyi, B., Mehta, S., & Maurer, A. (2019). Ordering of diagnostic imaging by physical therapists: A 5-year retrospective practice analysis. *Physical Therapy, 99*(8), 1020–1026.

Kendi, I. X. (2017). *Stamped from the beginning: The definitive history of racist ideas in America*. Bold Type Books.

Kesteloot, C., Loopmans, M., & De Decker, P. (2016). Space in sociology: An exploration of a difficult conception. In K. De Boyser, D. C. Dierckx, & J. Firedirch (Eds.), *Between the social and the spatial: Exploring the multiple dimensions of poverty and social exclusion* (pp. 113–132). Routledge (Chapter 7).

Lacroix, P. (2016). From strangers to "humanity first": Canadian social democracy and immigration policy, 1932-1961. *Canadian Journal of History*, 53–82.

Lincoln, Y. S., Lynham, S. A., & Guba, E. G. (2011). Paradigmatic controversies, contradictions, and emerging confluences, revisited. In N. K. Denzin, & Y. S. Lincoln (Eds.), *The sage handbook of qualitative research* (pp. 97–128). Sage.

Lorenzo, F. M. E., Dela Rosa, J. F., Paraso, G. R., Villegas, S., Isaac, C., Yabes, J. N., Trinidad, F., Fernando, G., & Tienza, J. (2005). *Migration of health workers: Country case study Philippines (International Labour Organization Working Paper Number 236)*. https://www.ilo.org/wcmsp5/groups/public/- - -ed_dialogue/- - -sector/documents/publication/wcms_161163.pdf.

Lurch, S. (2021). Stories of the visible and invisible: A 100 year history of movement. In *Canadian Physiotherapy Association Congress 2021*. https://www.embodiaapp.com/courses/1115-closing-keynote-presentation-stories-of-the-visible-and-invisible-a-100-year-history-of-movement-with-stephanie-lurch-canadian-physiotherapy-association.

Maldonado-Torres, N. (2007). On the coloniality of being. *Cultural Studies*, *21*(2), 240–270.

Marshall, S. C. (2022). *The history of physiotherapy guides its future*. Embodia. https://embodiaapp.com/courses/1472-the-history-of-physiotherapy-guides-its-future-canadian-physiotherapy-association.

Matthews, N. D., Rowley, K. M., Dusing, S. C., Krause, L., Yamaguchi, N., & Gordon, J. (2021). Beyond a statement of support: Changing the culture of equity, diversity, and inclusion in physical therapy. *Physical Therapy*, *101*, 1–5. https://doi.org/10.1093/ptj/pzab212.

Mignolo, W. (2011). *The darker side of Western modernity: Global futures, decolonial options*. Duke University Press.

Moffat, M. (2012). A history of physical therapist education around the world. *Journal, Physical Therapy Education*. https://journals.lww.com/jopte/Fulltext/2012/10000/A_History_of_Physical_Therapist_Education_Around.5.aspx.

Naidu, T. (2021). Modern medicine is a colonial artifact: Introducing decoloniality to medical education research. *Academic Medicine*, *96*(11S), S9–s12.

National Collaborating Centre for Determinants of Health. (2020). *Resource library. Let's Talk: Whiteness and health equity*. National Collaborating Centre for Determinants of Health. https://nccdh.ca/resources/entry/lets-talk-whiteness-and-health-equity.

Ndlovu, M. (2018). Coloniality of knowledge and the challenge of creating African futures. *Ufahamu: A Journal of African Studies*. https://escholarship.org/uc/item/7xf4w6v7.

Nixon, S. A. (2019). The coin model of privilege and critical allyship: Implications for health. *BMC Public Health*, *19*, 1637–1649. https://doi.org/10.1186/s12889-019-7884-9.

Nixon, S. A., Cleaver, S., Stevens, M., Hard, J., & Landry, M. D. (2010). The role of physical therapists in natural disasters: What can we learn from the earthquake in Haiti? *Physiotherapy Canada*, *62*(3), 167–171.

Ostry, J. D., Loungani, P., & Furceri, D. (2016). Neoliberalism: Oversold? *Finance & Development*, *53*(2), 38–41.

Paton, M., Naidu, T., Wyatt, T. R., Oni, O., Lorello, G. R., Najeeb, U., … Kuper, A. (2020). Dismantling the master's house: New ways of knowing for equity and social justice in health professions education. *Advances in Health Sciences Education*, *25*, 1107–1126.

Physiotherapy Education Accreditation Canada. (2020). *Accreditation standards for Canadian entry-to-practice physiotherapy*. https://peac-aepc.ca/pdfs/Accreditation/Accreditation%20Standards/Accreditation-Standards-for-Canadian-Entry-to-Practice-Physiotherapy-Education-Programs-(2020).pdf.

Quijano, A. (2000). Coloniality of power and Eurocentrism in Latin America. *Nepantla: Views from South*, *1*(3), 533–580.

Robinson, C. J. (2000). *Black Marxism: The making of the black radical tradition*. The University of Carolina Press.

Shah, T. I., Milosavljevic, S., Trask, C., & Bath, B. (2019). Mapping physiotherapy use in Canada in relation to physiotherapist distribution. *Physiotherapy Canada, 71*(3), 213–219.

Steinberg, S. (2005). Understanding the functionality of white supremacy. In L. Karumanchery (Ed.), *Engaging equity: New perspectives on anti-racist education* (pp. 13–26). Detselig Enterprises.

The Canadian Council of Physiotherapy University Programs/Le Conseil canadien des programmes universitaires de physiothérapie. (2023). *Canadian programs*. https://www.physiotherapyeducation.ca/programs.php.

Turpel-Lafond, M. E. (2020). *In plain sight: Addressing Indigenous-specific racism and discrimination in BC health care*. Government of British Columbia.

Vader, K., Ashcroft, R., Bath, B., Decary, S., & Deslauriers, S. (2022). Physiotherapy practice in Primary Health Care: A survey of physiotherapists in team-based primary care organizations in Ontario. *Physiotherapy Canada. Physiotherapie Canada*. https://pubmed.ncbi.nlm.nih.gov/35185252/.

Van Vuuren, M., & Westerhof, G. J. (2015). Identity as "knowing your place": The narrative construction of space in a healthcare profession. *Journal of Health Psychology, 20*(3), 326–337.

Vazir, S., Newman, K., Kispal, L., Morin, A. E., Mu, Y., Smith, M., & Nixon, S. (2019). Perspectives of racialized physiotherapists in Canada on their experiences with racism in the physiotherapy profession. *Physiotherapy Canada, 71*(4), 335–345.

Warf, B. (2010). *Encyclopedia of geography*. Sage Knowledge. https://sk.sagepub.com/reference/geography/n974.xml.

Wegrzyn, P., Judge, M., Lu, R., Smith, M., & Nixon, S. A. (2021). *A reorientation of belief: Considerations for increasing the recruitment of Black students into Canadian physiotherapy programs*. https://www.jhrehab.org/wp-content/uploads/2021/10/Wegrzyn_et_al_A_Reorientation_of_Beliefs_PDF_Fall_2021.pdf.

World Confederation for Physical Therapy. (2019). *Advanced physical therapy practice: Policy statement*. https://world.physio/sites/default/files/2020-07/PS-2019-APTP.pdf.

World Physiotherapy. (2020). *Annual membership 2020 census*. https://world.physio/sites/default/files/2021-02/AMC2020-Global_0.pdf.

World Physiotherapy. (n.d.). Glossary. https://world.physio/resources/glossary.

Yeowell, G. (2013). 'Oh my gosh I'm going to have to undress': Potential barriers to greater ethnic diversity in the physiotherapy profession in the United Kingdom. *Physiotherapy, 99*, 323–327.

York University. (2022). *White supremacy*. https://www.yorku.ca/edu/unleading/systems-of-oppression/white-supremacy/.

Yoshida, K., Self, H., & Willis, H. (2016). Values and principles of Teaching Critical Disability Studies in a physical therapy curriculum: Reflections from a 25-year journey-part 1: Critical disability studies value framework. *Physiotherapy Canada, 69*(1), 6–13. https://pubmed.ncbi.nlm.nih.gov/27904231/.

Chapter 10

Voices unheard: A clarion call for transforming communication sciences and disorders

Danai Kasambira Fannin[a,b], Mariam M. Abdelaziz[a], Nidhi Mahendra[b], and Jairus-Joaquin Matthews[c]

[a]*Department of Communication Sciences and Disorders, North Carolina Central University, Durham, NC, United States,* [b]*Department of Head and Neck Surgery and Communication Sciences, Duke University School of Medicine, Durham, NC, United States,* [c]*Department of Counseling, Higher Education, and Speech Language Pathology, University of West Georgia, Carrollton, GA, United States*

Fight for the things that you care about. But do it in a way that will lead others to join you.

Justice Ruth Bader Ginsberg

State of workforce diversity in communication sciences and disorders

The limited diversity in the allied health workforce is a barrier to accessible, equitable, and high-quality healthcare for an increasingly diverse population (Salsberg et al., 2021). Audiologists and Speech-Language Pathologists (SLPs) are the most qualified providers for preventing, screening, assessing, and treating communication disorders, but the workforce is not reflective of client demographics.

Scope of practice and regulatory agencies

Audiologists' scope of practice includes hearing, vestibular (balance), and other related disorders (American Speech-Language-Hearing Association (ASHA), 2018). While the precise scope can differ based on state regulations, certifications or licensure, and individual expertise, common duties include (a) hearing evaluation and vestibular function; (b) hearing aid appraisal and fitting; (c) tinnitus evaluation and management; (d) aural rehabilitation; (e) newborn

Equity, Diversity, and Inclusion in Healthcare. https://doi.org/10.1016/B978-0-443-13251-3.00010-7

hearing screenings; and (f) cochlear implant evaluation for candidacy and management. The SLP scope encompasses services for (a) speech production; (b) language; (c) voice; (d) resonance; (e) fluency; (f) cognition; (g) swallowing; (h) hearing (with necessary audiology collaboration); (i) advocacy; and (j) counseling for person-centered care (ASHA, 2015). These clinicians work in healthcare, private practices, and schools, guided by either the American Academy of Audiology (AAA), or the American-Speech-Language-Hearing Association (ASHA). The AAA provides professional support, resources, and advocacy for Audiologists, all to advance the quality of hearing and vestibular care. Audiologists can also be certified by ASHA which supports SLPs, Audiologists, scientists, Speech-Language Pathology Assistants (SLPA), and Audiology Assistants (AA).

Training and certification processes for clinicians are governed by two subsidiaries of ASHA. The Council for Clinical Certification in Audiology and Speech-Language Pathology (CFCC), a semiautonomous credentialing body of ASHA, oversees the certification of clinicians awarding the CCC-A (Certificate of Clinical Competence in Audiology), CCC-SLP (Certificate of Clinical Competence in Speech-Language Pathology), Audiology Assistant Certification (C-AA), and Speech-Language Pathology Assistant Certification (C-SLPA). The CFCC develops criteria for certification, manages the certification process, and administers the certification maintenance program. The Council for Academic Accreditation (CAA) establishes university accreditation standards and facilitates program improvement. The CAA can play an important role in workforce diversification, as accreditation evaluation includes indexes such as student body demographic profiles, admission and dismissal procedures, retention rates, the number of students taking longer than expected to graduate, Praxis exam (a national measure of knowledge and skills needed to prepare for working in educational settings) passing rates, and number of students securing a job within one year of graduation. Moreover, syllabi are reviewed for how multicultural aspects of client care are incorporated into didactic courses and clinical practicum. Though ASHA certification is not required by all employers, state licensure is, and state boards coordinate licensure and regulation of SLPs and Audiologists, with the responsibility to safeguard the public from malpractice.

Preprofessional requirements

In 2012, the entry-level degree for an Audiologist shifted from a Master's degree to a Doctor of Audiology (AuD), while the Master's degree has been the minimum requirement for SLPs since the late 1960s (ASHA, 2023a). For those who do not complete an undergraduate degree in communication disorders, such "post baccalaureate" students can take prerequisite courses before enrolling in graduate school. AAs and SLPAs recently became certified by ASHA, with options for the educational entry level including an Associate's degree, a Bachelor's degree, and even a certificate based on military job service and training for the AA. These Assistants work under the supervision of Audiologists and SLPs, aiding with assessment and intervention, which increases the availability of services. Although not licensed or certified by an association, the Doctor of Speech-Language Pathology (SLPD) is another advanced degree beyond the Masters focused on enhancing clinical expertise, leadership, and research capabilities. Thus, being an Audiologist or SLP requires graduate school, which poses a barrier to entry into the discipline. Moreover, this barrier is higher for those from low socioeconomic status (SES) households, given that low and middle SES high schoolers earn college degrees at significantly lower rates than high SES students (National Center for Education Statistics, 2015). Data from 2016 to 2020 from the Communication Sciences and Disorders Centralized Application Service (CSDCAS), an admissions platform used across allied professions, indicates that historically marginalized populations and those being the first generation to attend college were significantly less likely to receive acceptance offers (Watts et al., 2023). As such, critical reflection on the following admission requirements that can pose systemic barriers to entry for diverse applicants is needed.

Admission requirements

Of the 321 graduate SLP and 80 Audiology programs in the US (ASHA, 2023c), 67% participate in CSDCAS (Woods, 2022). As CSDCAS has the most readily available data, it will serve as an exemplar of admission trends for the purposes of this chapter. Whereas the minimum grade point average (GPA) tends to be 3.0 for graduate admission, the mean undergraduate GPA for accepted AuD students in 2021–22 was 3.59 (median = 3.67; standard deviation = 0.33), and was 3.60 (median = 3.66; standard deviation = 0.33) for SLP students. The mean GPA for denied AuD students and denied SLP students was 3.29 and 3.26, respectively (Woods, 2022), demonstrating the competitiveness of admission. Beyond GPA, the Graduate Record Examination (GRE) has been an important metric but was suspended during the pandemic years of 2020–22. Prepandemic, 95% of programs required the GRE but only 26% of AuD programs did in 2022–23, with 62% of the remaining programs not requiring it at all, and 12% making the GRE optional. For SLP programs, 17% required the GRE in the 2022–23 cycle, with 11% having it optional, and 72% not requiring GRE (Woods, 2022), showing that the measure is falling out of favor.

The remaining application components usually include a personal statement/essay and two to three recommendation letters, with the use of interviews varying widely across programs. Within these qualitative application components, programs are typically looking for evidence of volunteerism, leadership, community involvement, and research experience. These unspecified expectations favor privileged students with more free time, strong financial support, and higher-educated social networks than students who are not as privileged

(Capers IV, 2020; Mandulak, 2022). In fact, Ellis et al. (personal communication, May 31, 2023) conducted analyses showing that applicants from minoritized populations in the CSDCAS database tended to have extracurricular activities that were not as highly valued by admission reviewers, possibly explaining their receipt of lower application ratings. Moreover, recommendation letters are implicitly biased against applicants usually underrepresented in programs and the workforce (Newkirk-Turner & Hudson, 2022), thereby putting the onus of writing careful, unbiased letters on faculty members. University programs serve as the primary workforce pipeline, and the following section will outline diversity trends in the resultant workforce.

Current workforce diversity numbers

In 2022, the 214,654 members of ASHA experienced growth, with an increase of 6288 individuals, reflecting a 2.9% increase from 2011 (ASHA, 2023b). However, only 8.9% self-identified as individuals from underrepresented racial or ethnic groups, with 6.4% of those identifying as Hispanic or Latine (ASHA, 2023b), while 38.4% of the US residents belong to the underrepresented racial/ethnic groups (with 18.7% being Hispanic or Latine) (U.S. Census Bureau, 2020). In addition, 95.6% of ASHA members identify as female, showing a slight increase from 95.5% in 2020, despite initiatives to recruit more males. Specifically, 13.4% of Audiologists, 3.6% of SLPs, and 15.9% of those with dual certification were male while, comparatively, 49.5% of the US population were male. Because providers are concerned with speech and language disorders, as well as culturally and linguistically responsive care, multilingualism is especially important to our practitioners. At the end of 2022, 8.3% of ASHA members self-identified as multilingual—up slightly from 8.2% in 2021, with 86 languages represented, including sign language (ASHA, 2023b). Percentages such as 8.3%–8.7%, however, remain concerningly low when compared to the 21% of the US population over 5-years-old that speak a language other than English (U.S. Census Bureau, 2020).

Having diverse representation among providers is essential for delivering optimal and equitable treatment and outcomes. The national professional organizations such as AAA, ASHA, and the Council of Academic Programs in Communication Sciences and Disorders (CAPCSD) recognize the cultural-linguistic incongruence between the workforce and clients. Indeed, ASHA's Office of Multicultural Affairs and ASHA members have addressed this issue for over 53 years; yet, progress has been gradual. This slow progress is reflected in ASHA's demographic profile, and in a special forum of the *American Journal of Speech Language Pathology* published in 2021 that highlighted trends and approaches to diversifying the workforce (Kasambira Fannin & Mandulak, 2021). This chapter will outline (1) trends in diversification for the discipline, (2) a rationale for systemic action to diversify the workforce, and (3) suggestions on how to better diversify the workforce.

Longitudinal diversity trends in the communication sciences and disorders workforce

ASHA formed the Office of Multicultural Affairs in 1969 during the Civil Rights movement in response to the need for access to quality services for minoritized populations. Because progress in workforce diversification has been slow, a resurgence of calls for social justice and examining racism in the discipline occurred during the Black Lives Matter movement which peaked in 2020 (Deal-Williams, 2020; Kasambira Fannin, 2020; Mahendra et al., 2021; Mahendra & Visconti, 2021). A longitudinal report of ASHA demographic trends from 2002 to 2022 illustrates that, in two decades, male representation has decreased, and racial/ethnic diversity has improved minimally. In 2022, 86.9% of ASHA Audiologists and 96.4% of SLPs identified as female (compared to 80.7% and 95.3% in 2002). In 2022, 8.9% of ASHA members were from underrepresented racial groups, compared to 4.9% in 2002, and approximately 6% reported Hispanic or Latine ethnicity in 2022 compared to 2.5% in 2002 (ASHA, 2023b). Ultimately, the lack of racial/ethnic diversity reached the public sphere in 2013 when an article in *The Atlantic* identified speech-language pathology as the *"fourth Whitest profession in America"* (Thompson, 2013). Decreasing male membership has also been a concern since the early to mid 1990s, sparking ASHA's initiatives to recruit males (Matthews & Daniels, 2019). Yet, as in other female-dominated disciplines, males are overrepresented among PhD recipients, awards of tenure and promotion, academic leadership, and recipients of marquis ASHA or AAA awards (Rogus-Pulia et al., 2018). Still, increased representation of males in the workforce would better reflect US demographics. Moreover, perhaps, more males joining the discipline might increase compensation for all clinicians, as the primary cause of the gender pay gap is whether an occupation is dominated by males or females, explaining more than 50% of the pay gap (Blau & Kahn, 2017; Deal-Williams, 2010).

Although diversification progress has been slow, a more fine-tuned measure has been developed to capture the progress that has been made, to promote momentum and accountability for increased diversification. The ASHA Diversity Index assesses changes in percentages of members who identify as belonging to one or more underrepresented groups in the association (ASHA, 2022). The purpose is to supplement the limited, annual raw demographic data with additional accounting for intersectional identities. The Diversity Index indicates a 3% increase in diversity from 2007 to 2021 but a 7% increase within the new ASHA member (5-year) cohort. This shows diversity increasing faster among newer members and, thus, focused recruiting efforts displaying their positive impact.

The rationale for systemic action and solutions to improve workforce diversity

Along with other systemic factors, the racial/ethnic mismatch between the clinician and client has a direct and significant impact on educational and health disparities (Salsberg et al., 2021; Yong-Hing & Khosa, 2023; Young et al., 2023). The next section will detail how a homogenous workforce can contribute to suboptimal care.

Educational disparities in communication sciences and disorders

Children from marginalized backgrounds are disproportionately represented in specialized education services including speech-language pathology and audiology (Office of Special Education and Rehabilitative Services, 2021; Robinson & Norton, 2019). Disproportionality, a measure of educational equity, refers to both the under- and overrepresentation of particular groups in special education via the federal Individuals with the Disabilities Education Act of 2004 (IDEA) areas of eligibility (e.g., specific learning disability, speech-language impairment). Black children are overrepresented in special education at 17%, while making up only 15% of the public school population (National Center for Education Statistics, 2023). However, White children make up 45% of the public school population but only 15% of the special education population (NCES, 2023). Similar overrepresentation is seen among American Indian/Alaska Native (general school population—1%; special education—19%), and Pacific Islander children (general school population less than—1%; special education—11%) due, in part, to the use of assessments that do not consider cultural and linguistic differences. Such differences often render standardized tools inappropriate and culturally invalid, yet their use persists.

Once assigned to special education, racial/ethnic differences in access to and provision of services emerge. For example, Black children are overrepresented in the IDEA eligibility category of severe emotional disturbance or intellectual disability and are less likely to receive services for speech and language (Kim et al., 2021; Morgan et al., 2017; Office of Special Education Programs, 2021a). Conversely, Black and Latine children tend to be underidentified or receive a later identification than their White counterparts for conditions such as autism, which is also medically diagnosed (Baio et al., 2018; Magaña et al., 2012; Morgan et al., 2017; Yingling et al., 2018), and underrepresentation of Latine children in special education, in general, also occurs (NCES, 2023).

Bias in educational assessment

Bias in special education referrals contributes to disproportionality and ensuing disparities in quality of life. Most referrals, around 75%, come from teachers rather than parents (New York University, 2016). Although the teaching profession is more diverse than communication disorders, the odds of a minoritized student having a teacher who shares their background are lower, compared to White students, as the teacher workforce was 80% White in 2020–21 (Schaeffer, 2021). Cultural incongruence may lead to inaccurate or missed referrals when students present in a way that does not align with a teacher's cultural perspective on typical behaviors or disability (Hyter & Salas-Provance, 2023). For instance, a teacher or SLP unfamiliar with African American or Chicano English may view statements like "I be running" or "I jump on the water" as incorrect, rather than grammatically correct dialectal characteristics. Fish (2019) found that teachers were more likely to identify academic difficulties as disabilities in White boys but not in Boys of Color facing the same challenges. This implies lower expectations and assumptions of lower performance levels for Boys of Color. Consequently, a homogenous workforce failing to mitigate implicit bias and not recognizing and adjusting one's own cultural lens affects the accuracy of referrals and placement for diverse students.

Disparities in school disciplinary practices

Research shows that students from marginalized backgrounds face a higher likelihood of out-of-school suspensions and expulsions (Jones et al., 2018; Stanford, 2019). This disproportionality is amplified for children in special education (Stanford, 2020). Specifically, Black children with language and/or learning disabilities are twice as likely to be suspended compared to White children (Stanford, 2020) and approximately 45% of youth with communication disorders are involved in the criminal justice system (Stanford, 2019). The implementation of zero tolerance policies contributes to this issue and supplies the school-to-confinement pipeline (American Civil Liberties Union, 2022). As such, having more culturally congruent or responsive clinicians who recognize the link between communication disorders, race/ethnicity, SES, and biased disciplinary actions could reduce the high representation of disabled children involved in the criminal justice system. Disproportionality in eligibility categorization and discipline is a result of multiple complementary and supplementary factors: (1) inequitable access to resources; (2) culturally inappropriate evaluation; (3) lack of community outreach, education, and collaboration; and (4) implicit and explicit bias. Each of these factors is also tied to health disparities.

Health disparities in communication disorders

Chinn et al. (2021) discovered persistent healthcare disparities at the intersection of race/ethnicity and social determinants of health (see Fig. 1) (U.S. Department of Health and Human Services, 2022). Stroke, a primary reason for SLPs to see clients, disproportionately affects Black individuals (Davis et al., 2014) and is the leading cause of disability in the US (Harris et al., 2006; Howard et al., 2005). Non-Hispanic Black US residents have higher rates of elevated cholesterol, Type II diabetes, heart disease, and hypertension, doubling their risk of stroke and dementia (Howard & Howard, 2020; Mozaffarian et al., 2016; Yong-Hing & Khosa, 2023). While stroke prevention and treatment

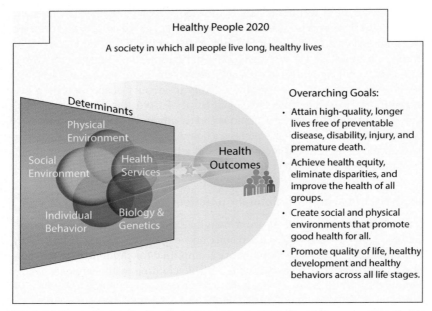

FIG. 1 A figure showing the five Social Determinants of Health and the goals of the Healthy People 2020 initiative to improve health outcomes.

efforts have reduced mortality (Go et al., 2014; Lackland et al., 2014), chronic disabilities such as swallowing and communication disorders persist (Winstein et al., 2016). Communication disorder caseloads are more diverse than the current demographic of 93% White female clinicians. When there is a racial/ethnic mismatch between patients and providers, some patients may mistrust providers, resulting in less effective therapeutic relationships and increased disparities in outcomes (Jetty et al., 2022; Mahendra & Spicer, 2014; Morgan et al., 2017; Penner et al., 2010; Yong-Hing & Khosa, 2023).

The following section includes a review of factors adversely influencing treatment outcomes and impeding equitable, quality care.

Inappropriate assessment practices in education and healthcare

Many school districts have adopted standardized testing policies for special education eligibility, despite it not being an IDEA requirement (O'Malley, 2015). These measures are often influenced by SES rather than genuine ability (Finneran et al., 2020). Instead, culturally sensitive approaches such as dynamic assessment (Shipley & McAfee, 2019) and language sampling (Fulcher-Rood et al., 2018) are recommended, although they can be time-consuming. It is important to note that alternative measures alone are insufficient, as the data require interpretation by clinicians who bring their diverse lived experiences and perspectives (Donini-Lenhoff & Hedrick, 2000; Hood et al., 2015).

Diagnostic errors can occur when clinicians rely on heuristics (i.e., stereo-types) for immediate diagnoses, despite their limitations (Newman-Toker, 2014; Norman & Eva, 2010). Clinicians' perceptions of demographic groups can be influenced by their experiences, training, and practice settings, leading to the formation of stereotypes that manifest as diagnostic heuristics (Branch Jr., 1998; Donini-Lenhoff & Hedrick, 2000). In the context of communication dis-orders, clinical training programs often prioritize exposing students to patients from low SES and other diverse groups (Betancourt, 2004). This can result in clinicians associating certain groups with specific health behaviors or beliefs, which may be influenced more by social determinants such as poverty or the physical environment rather than their cultural background (Donini-Lenhoff & Hedrick, 2000). Consequently, clinicians may inaccurately apply heuristics based on their clinical practicum experiences, leading to erroneous diagnostic and treatment decisions (Chin, 2000). For instance, if a clinician assumes that all Black people distrust researchers due to a history of unethical medical exper-imentation, they might fail to offer Black patients the opportunity to participate in a clinical trial. Such disparity in clinical trial recruitment has been recently documented in several publications (Ding et al., 2021; Tao et al., 2023). Addi-tionally, cultural factors, including dialect use, can influence clinicians' percep-tion of dysarthric speech (Dagenais & Stallworth, 2014). Research by Perrachione et al. (2010) demonstrated that listeners rated speech intelligibility and comprehensibility higher for speakers of their own race, with White lis-teners giving significantly higher ratings to White dysarthric patients who did not use dialects. These findings reveal an implicit bias favoring one's own race and language use, possibly stemming from clinicians' limited famil-iarity with patients who speak dialects. Overall, clinical decision-making is neg-atively impacted by implicit biases, flawed heuristics stemming from well-intended exposure to diverse settings during training (Betancourt, 2004), insufficient training to control implicit bias among clinicians, and a lack of diversity within the healthcare workforce (Ding et al., 2023). Addressing these issues requires both increasing workforce diversity and equipping all cli-nicians with cultural responsiveness training to ensure equitable services for all.

Strategies to diversify the workforce

We have reviewed the far-reaching impacts of having a workforce that is not in congruence with US demographics and documented some reasons for these issues. A growing body of research demonstrates the advantages of students and patients receiving services from providers who share a similar cultural background. Tangible benefits of Equity, Diversity and Inclusion (EDI) are (a) improved workforce morale, (b) reduced employee burnout and turnover, (c) provision of mentors and role models for minoritized students and, (d) provision of culturally competent care (Ding et al., 2023; Yong-Hing & Khosa, 2023; Yong-Hing et al., 2023; Young et al., 2023). When there is a racial/ethnic "match" between educators and students, academic benefits include: (1) more Black and Latine students in gifted programs (ASHA, 2018; Grissom et al., 2017), (2) better performance on math and reading evaluations (Dee, 2004; Ouazad, 2014), and (3) a decrease in chronic absenteeism (Holt & Gershenson, 2015). Provider-patient racial/ethnic congruence correlates with decreased mortality (Alsan et al., 2021; Hill et al., 2023) and increased patient satisfaction (Howard et al., 2001).

However, 93% of ASHA members are monolingual English speakers, and research shows that most SLPs in the US receive education primarily focused on monolingual English language development (Alfano et al., 2021; Durán et al., 2016; Getzler et al., 2021; Paz et al., 2023). Furthermore, studies show that most SLPs feel inadequately prepared to assess and provide treatment for multilingual clients (Núñez & Hughes, 2021). Increasing language diversity among clinicians is, therefore, one reason to diversify the workforce (Alfano et al., 2021). Parveen and Santhanam (2021) demonstrated the benefit of language diversification where bilingual SLPs reported higher competency working with non-English speakers, compared to monolingual SLPs. This bilingual SLP advantage was further supported by Narayanan and Ramsdell (2022) who observed that multilingual SLPs reported a greater growth in their understanding of concepts and confidence working with CLD clients, than their monolingual peers.

Several initiatives have been proposed to address the need for more diverse practitioners (Langford-Hall, 2016). These include strategic recruitment and hiring of more underrepresented faculty (Horton et al., 2021), ASHA's Minority Student Leadership Program, allocating grant funds for research on multicultural issues or bilingual personnel preparation, and integrating inclusive, equitable practices in admissions, advising, instruction, clinical supervision, and mentoring, as part of ASHA's Strategic Objectives. Specific groups amongst ASHA members have contributed notably to these efforts. For example, ASHA Special Interest Groups (SIG) 14 (CLD) and SIG 17 (Global Issues in Communication Sciences and Related Disorders), L'GASP: The LGBTQ caucus of ASHA, as well as online communities such as the Allied Health Specialists with Disabilities Alliance, promotes diversity beyond race/ethnicity. Expansion of the definition of diversity is exemplified in Abdelaziz et al. (2021) study of student microaggressions where 10 different categories (e.g., transgender, multilingual, disabled, and nontraditional student) were identified to represent the types of clinicians we need to better represent in the workforce. The following section includes ideas to increase workforce diversity.

Preprofessional EDI initiatives

It is essential to integrate EDI principles into the CSD curriculum. EDI should be woven throughout the curriculum and explicitly taught, to instill these principles in students to carry to future work environments where they will supervise employees and graduate clinicians. In fact, several ASHA members have called for the CAA to require a stand-alone course on multicultural aspects of communication disorders, as suggested by Stockman et al. (2008), who found that instructors who taught stand-alone multicultural courses ($N = 731$) reported more positive instructional outcomes compared to those with only multiculturalism infused classes. Preprofessional EDI training programs are being implemented to ensure that students graduate with knowledge and an inclusive mindset towards their current and future colleagues. Students who experience marginalization, exclusion, or discrimination from teachers or peers may face increased stress and mental health issues, putting them at a higher risk of withdrawing from these programs (Abdelaziz et al., 2021; Deal-Williams, 2020). Therefore, it is crucial to create inclusive educational environments that support underrepresented students and promote their retention. Affinity groups have been a resource for constructive discussions and collective action on issues affecting minoritized students (Alicea & Johnson, 2021; Brea-Spahn, 2021; Myers et al., 2019; Scott, 2023). Starting affinity groups at the undergraduate level is important, given the historical underrepresentation of minoritized students in graduate programs. These groups can provide undergraduates with a supportive environment, offering insights into graduate school, tutoring and mentoring by graduate students, and guidance throughout the application process.

Affinity groups also play a vital role for graduate students, offering support for academics and clinical responsibilities. At a national level, ASHA recognizes the need to address systemic racism and oppression faced by students, and the National Student Speech-Language Hearing Association (NSSLHA) is actively working towards increasing cultural responsivity, cultural humility, and cultural sensitivity through programming and collaboration with multicultural constituency groups representing intersectional members. For example, NSSLHA organized a webinar on choosing an inclusive graduate program, where faculty panelists agreed to serve as objective advisors or advocates for students facing problems. This suggestion was prompted by Abdelaziz et al. (2021) study on microaggressions, revealing that student grievances were dismissed or ignored by faculty, supervisors, and administration, and students feared retaliation if they reported a concern.

Professional DEI initiatives

CAPCSD and ASHA offer EDI-focused continuing education, including the Microaggression Microcourse Series and holistic admissions training. Several of these offerings are free, aiming to highlight the importance of diversity in student and workforce populations. Workplaces now provide EDI training through Grand Rounds discussions among hospital SLPs and Audiologists, as well as discussions on EDI-related journal articles/lectures, inclusive supervision of graduate clinicians, and mitigating implicit bias in staff interviews

(J. Minga, personal communication, February 1, 2023). The CAA introduced a multicultural continuing education requirement for certification renewal.

Creating inclusive environments should not be solely the responsibility of faculty and students. Universities must actively involve preceptors in promoting inclusivity, as students spend significant time with them during clinical practicums, which can also serve as informal job interviews. Professional expectations apply to everyone in all work settings, and ASHA's Code of Ethics mandates professional conduct to safeguard clients, research participants, and the profession's reputation (ASHA, 2023d). Preceptors can impact students through grading for professionalism, an ambiguous practice rife with implicit bias. Definitions of professionalism typically encompass behavior, appearance, and personal traits that supposedly indicate job performance, but they are often vague and vary across professions (Bhatia-Lin et al., 2021). Consequently, professionalism can become a tool for control by dominant social groups within the discipline (Frye et al., 2020), with measures of professionalism often aligning with the standards of a typical communication disorders professional (e.g., middle socioeconomic status, White, monolingual English-speaking, American, cisgender female without reported disabilities).

> *Professionalism has become a coded language for white favoritism in workplace practices that more often than not privilege the values of white and Western employees and leave behind people of color.*
>
> Gray (2019)

Thus, without unbiased grading rubrics for professionalism, structural racism and ableism will manifest (Porter, 1993), allowing supervisors to marginalize students under the pretense of professional requirements. Recognition of these issues and the use of unbiased grading criteria should therefore be included in EDI training for faculty, preceptors, and employers. If a preceptor is found to be harmful to a student, there must be real recourse, including the university clinic director discussing incidents or grading anomalies with the preceptor, advocating for the student, and ultimately replacing the preceptor if there is a risk of harm to present or future students. A proactive way to approach equitable grading of "soft skills" such as professionalism could be to offer continuing education in being effective supervisors for underrepresented students and also having clear, written guidelines on the program's expectations of preceptors. Most importantly, programs should swiftly investigate any concerns students report about preceptors, as should workplace administrators when employee concerns arise.

Increased funding

Public schools rely on local, state, and federal funds but, despite being a federal law, the Individuals with Disabilities Education Act (IDEA) has not received full funding from Congress. Only 14% of the promised 40% funding has been provided (National Center for Learning Disabilities, 2022), despite a 25% increase in students served under IDEA over the past 20 years. Increased funding could trim caseloads, allowing for more thorough evaluations using least-biased measures instead of sole reliance on standardized tools that may be efficient but less accurate. Moreover, it is important to consider specific funding for bilingual providers, given the added workload involved in bilingual evaluations. Premium pay could be a recruitment incentive for multilinguals who are often underrepresented (Kritikos, 2003), yet interested in working in allied health professions.

ASHA and AAA allocate resources to advocate for relevant public legislation, such as loan repayment programs for school clinicians or the Allied Health Workforce Diversity Act (H.R. 3320/S. 1679) which became law in December 2022. This legislation enables the Department of Health and Human Services to offer grants to enhance diversity in physical therapy, occupational therapy, respiratory therapy, audiology, and speech-language pathology through recruitment and retention scholarships. Members of other national allied health associations also played a crucial role in sustaining the legislation's momentum by advocating in their home districts and on Capitol Hill.

Equitable and inclusive admissions processes

To address biases in programs, adopting holistic admissions is imperative. This approach dismantles systemic gatekeeping and promotes diversity. Programs that embrace holistic admissions have more diverse cohorts, increased opportunities for first-generation students, and faculty recognition of the value of applicant diversity. Communication disorders programs should therefore reassess admission processes and include diversity of life experience as a core evaluation criterion (Guiberson & Vigil, 2021; Mandulak, 2022; Wong et al., 2021).

The holistic review evaluates applicants' experiences, attributes, and academic achievements while considering their potential contributions to the profession and program (Guiberson & Vigil, 2021). One strategy would be to eliminate or make the GRE optional, as the GRE does not reliably predict success in graduate school (Petersen et al., 2018; Sealy et al., 2019). This change would address the leaky pipeline between undergraduate and graduate programs, where universities find few of their own undergraduates gaining admission (often due to lower GRE scores). Even at Historically Black Colleges and Universities (HBCUs), non-Black students made up 25% of enrollment in 2022 (OSEP, 2021b), compared to 15% in 1976 (NCES, 2022). Some HBCU communication disorder programs have had up to 75% White students, exhibiting how graduate students at minority-serving institutions can still be majority White females, reflective of their 91% proportion among the ASHA membership (ASHA, 2023b). For programs that still require the GRE, adopting a holistic admissions strategy would involve having faculty review other application components without the knowledge of GRE scores. This blind review process helps mitigate implicit biases associated with lower GRE scores, preventing these biases from influencing scores of unrelated components such as interviews and personal statements.

Holistic admissions have been used in medicine longer than in communication disorders; thus, implementation is not yet widespread (Capers IV, 2020). Few longitudinal analyses exist on the outcomes of holistic admissions in communication disorders, with one study showing that, during a holistic admissions simulation, respondents ($N = 66$) were still less likely to make an explicit "accept" decision for applicants who varied from the "high achieving" stereotyped candidate (Girolamo et al., 2022). To address the gap in the knowledge base, Mandulak and Kasambira Fannin (2023) conducted an admissions workshop for 25 programs at the 2023 CAPCSD annual convention. Participants from this 2023 workshop will collaborate over the next three years with 25 programs from the 2024 cohort to collect and analyze data for changes in their class composition after implementing holistic admissions principles. The workshop and subsequent analyses will be repeated biannually at future conventions.

Marketing

We recognize the challenge of promoting awareness about speech-language pathology and audiology, as they are niche disciplines. Given the nationwide shortage of clinicians, proactive recruitment efforts are essential and early recruitment from middle and high schools and undergraduates from other majors can ultimately enhance workforce diversity. Richburg (2022) highlighted the limited exposure of Black college students and males to the profession, and how organizations are addressing this problem. For example, members of the National Black Association for Speech-Language and Hearing (NBASLH) visit high schools annually and the National Student Speech-Language-Hearing Association leverages state-level Student Officer leaders to recruit high school students. For 50 years, ASHA has led an annual campaign in May called *National Speech-Language-Hearing Month*, which has included public service announcements, bilingual media tours, digital campaigns, and downloadable resources and information for the public and professionals focused on speech and hearing.

ASHA's Office of Multicultural Affairs developed promotional materials to support clinicians in community outreach efforts. These materials include race/ethnicity-specific pamphlets, video clips featuring clinicians and researchers, slide presentations about the discipline, and posters showcasing SLPs and Audiologists from underrepresented backgrounds. CAPCSD's Admissions Committee created templates to simplify the process of obtaining permission to present at organizations and schools. They also provided ideas for grade-level appropriate speech and hearing activities. Guides for parents and community college academic advisors were also developed to keep them informed about prerequisite requirements and academic opportunities for interested students.

Reevaluation of program policies

Systemic and programmatic policies require revision at both macro- and micro-levels. For instance, at the macrolevel, the CAA should reconsider penalizing programs based on graduation timelines, which unintentionally discourage the admission of students with potential challenges. Additionally, programs may dismiss students (and not readmit them upon appeal) if they anticipate the student needing to take nine credits instead of 12 per semester. Making changes in national accreditation policies that affect program-level decisions could increase diversification.

At the microlevel, the use of Essential Functions, derived from Human Resources practices, can be misapplied in educational programs. In communication disorders, it has been wrongly utilized to dismiss students with disabilities or differences, even when the prescribed "required skills" can be accommodated or are unnecessary for being an Audiologist or SLP. Essential Functions have also hindered student admissions if applicants disclose invisible disabilities (e.g., ADHD, anxiety, and depression), as programs may claim it could impair their ability to fulfil the tasks listed. In light of recent examination of how Essential Functions can disadvantage minoritized students in the discipline (Yu et al., 2022), more communication disorders programs are foregoing the use of the guidelines.

Revising strict undergraduate grade requirements can improve admission rates for applicants with diverse backgrounds and experiences. For instance, general education courses (e.g., Biology, Physics/Chemistry, and Statistics) that are not necessarily a predictor of clinical skills are sometimes used to deny admission, even though CFCC considers "passing" to be a D or above. The requirement for high grades in all classes, even when the classes are not relevant to the discipline, is an unnecessary obstacle to admission. Some programs consider every undergraduate grade from when the student was a new high school graduate. Young students, especially those who are the first to attend college in their families, are more prone to early academic missteps. Students from low SES backgrounds may also attend community colleges where they might encounter a lack of academic and career guidance. Some programs therefore only analyze grades from the latter two years of coursework when students should be more acclimated to college life, with improved study skills.

When considering relevant prerequisite courses (e.g., Language Development, Anatomy and Physiology of the Speech and Hearing Mechanism, and Phonetics), some programs require the applicant to retake courses with a C- or even B-grade, while other programs allow as low as "passing" D. Sometimes, the applicant or recommenders explicitly state why a student did not do well in a class (or GRE). When they have not explicated the lower grades, however, holistic admissions would have reviewers determine if tacit information in recommendation letters, personal statements, interviews, or societal events (such as the COVID pandemic) might explain a time when the student had lower grades.

Other policies that prohibit retention of nontraditional, caregiving, or working students include not allowing part time attendance, not permitting a student to take fewer courses at a time, or prohibiting incomplete grades. While universities allow course load changes under extenuating circumstances, rigid programs aiming for 100% graduation in two years may resist accommodating such requests. This can lead to stressful battles for accommodations, resulting in suboptimal educational experiences. Hence, flexibility in how courses are completed can improve student retention.

A significant amount of professional education includes clinical practicum. Prioritizing domestic students over international students for clinical practicum placements inhibits an inclusive environment. Furthermore, assigning students with different accents or dialects to less desirable sites (e.g., distant or not related to the student's interest) due to the preconceived notion that their speech or language differences would make them less qualified is discriminatory. The CAA considers such practices to be serious violations of accreditation standards. Additionally, scheduling classes that pose difficulties for students commuting from distant clinical sites using public transportation engenders a homogenous and stressed cohort. Rather, embracing instructional technology for asynchronous online courses during full-time external clinical placements can enhance student well-being. Careful introspection of program/workplace policies and openness to change is fundamental to fostering student diversity and creating a more inclusive workforce. To conclude, the following section lists resources that have been used to support the diversification of the communication disorders workforce.

AAA Supporting Diversity in Higher Education:
https://www.audiology.org/about/diversity-equity-and-inclusion/supporting-diversity-in-higher-education/

AAA Supporting Diversity in the Workplace:
https://www.audiology.org/about/diversity-equity-and-inclusion/supporting-diversity-in-the-workplace/

AAA Supporting Diversity in Healthcare Settings:
https://www.audiology.org/about/diversity-equity-and-inclusion/supporting-diversity-in-health-care-settings/

ASHA Member Advocacy:
https://www.asha.org/advocacy/asha-member-advocacy/

ASHA Career Toolkit and Recruitment Materials:
https://www.asha.org/careers/career-guidance-tools/

ASHA Office of Multicultural Affairs Resources:
https://www.asha.org/practice/multicultural/

ASHA Multicultural Constituency Groups:
- Asian Pacific Islander Speech-Language-Hearing Caucus: https://apislhc.org/

- Disability Caucus: SLPdisabilitycaucus@gmail.com
- Haitian Caucus: https://www.hcasha.org/
- Hispanic Caucus: https://www.hispaniccaucusasha.org/
- L'GASP—LGBTQ Caucus: https://lgbtqcsdsa.org/professionals/lgasp/
- Middle East and North Africa Caucus: MENAMCCG@gmail.com
- National Black Association for Speech-Language & Hearing (NBASLH): http://www.nbaslh.org/
- Native American Caucus: https://sites.google.com/view/nacslpa-org/home
- South Asian Caucus: http://www.sac-asha.org/

Council on Academic Accreditation in Audiology and Speech-Language Pathology:
  https://caa.asha.org/

Council of Academic Programs in Communication Sciences and Disorders:
  https://www.capcsd.org/

Select Cultural Inclusivity Resources Appropriate For Allied Health Professionals:

- Microaggressions Micro Course Series https://www.asha.org/practice/multicultural/microaggressions-micro-course-series/
- Supporting and Working With Transgender and Gender-Diverse People https://www.asha.org/practice/multicultural/supporting-and-working-with-transgender-and-gender-diverse-individuals/
- Cultural Competence Check-Ins Self-Assessment https://www.asha.org/practice/multicultural/self/

Strategies to Reduce/Neutralize Implicit Bias in Admissions Interviews (compiled by Quinn Capers, IV, MD, FACC)
  https://www.aamc.org/about-us/mission-areas/medical-education/interviews-gme-where-do-we-go-here#recommendation1

Acknowledgments

We would like to thank all the volunteer leaders we have worked with and colleagues nationwide who are committed to diversifying the field and advocating for inclusive educational environments for all of our students.

Disclosures

All the authors are paid salary by their respective primary institutions. The authors have been members of their universities' graduate admissions committees,[a,b,c,d] the American-Speech Language-Hearing Association (ASHA) Multicultural Issues Board,[a,b,c,d] the Coordinating Committee for ASHA's Special Interest Group 14 (Cultural and Linguistic Diversity) and Editors for its journal,[a,c] the Council of Academic Programs in Communication Sciences and Disorders (CAPCSD) Diversity Equity and Inclusion Committee,[a,c] the CAPCSD Student Recruitment Committee,[a] and coordinator of the CAPCSD Admissions Summit.[a]

References

Abdelaziz, M. M., Matthews, J. J., Campos, I., Kasambira Fannin, D., Rivera Perez, J. F., Wilhite, M., & Williams, R. M. (2021). Student stories: Microaggressions in communication sciences and disorders. *American Journal of Speech-Language Pathology, 30*(5), 1990–2002.

Alfano, A. R., Medina, A. M., & Moore, S. (2021). Preparing culturally and linguistically diverse students to work with culturally and linguistically diverse populations: A program design and student outcomes study. *Teaching and Learning in Communication Sciences and Disorders, 5*(3), 2.

Alicea, C. C., & Johnson, R. E. (2021). Creating community through affinity groups for minority students in communication sciences and disorders. *American Journal of Speech-Language Pathology, 30*(5), 2028–2031.

Alsan, M., Stanford, F. C., Banerjee, A., Breza, E., Chandrasekhar, A. G., Eichmeyer, S., Goldsmith-Pinkham, P., Ogbu-Nwobodo, L., Olken, B. A., Torres, C., Sankar, A., Vautrey, P., & Duflo, E. A. (2021). Comparison of knowledge and information-seeking behavior after general COVID-19 public health messages and messages tailored for Black and Latinx communities: A randomized controlled trial. *Annals of Internal Medicine, 174*(4), 484–492.

American Civil Liberties Union. (2022). *School-to-prison pipeline*. Retrieved May 6, 2023 from https://www.aclu.org/issues/juvenile-justice/juvenile-justice-school-prison-pipeline.

American Speech-Language-Hearing Association. (2015). *Scope of practice in speech-language pathology*. Retrieved February 2, 2023 from https://www.asha.org/policy/sp2016-00343/.

American Speech-Language-Hearing Association. (2018). *Scope of practice in audiology*. Retrieved February 22, 2023 from https://www.asha.org/policy/sp2018-00353/.

American Speech-Language-Hearing Association. (2022). *Diversity in ASHA rises*. Retrieved May 1, 2023 from https://leader.pubs.asha.org/do/10.1044/leader.AAG.27072022.diversity-membership.20/full/.

American Speech-Language-Hearing Association. (2023a). *A chronology of changes in ASHA's certification standards*. Retrieved April 2, 2023 from https://www.asha.org/certification/ccc_history/.

American Speech-Language-Hearing Association. (2023b). *Member and affiliate profile trends: 2002-2022*. Retrieved May 1, 2023 from https://www.asha.org/siteassets/surveys/2002-2022-member-and-affiliate-profile-trends.pdf.

American Speech-Language-Hearing Association. (2023c). *EdFind*. Retrieved April 2, 2023 from https://find.asha.org/ed/#sort=relevancy.

American Speech-Language-Hearing Association. (2023d). *Code of ethics*. Retrieved March 22, 2023 from www.asha.org/policy/.

Baio, J., Wiggins, L., Christensen, D. L., Maenner, M. J., Daniels, J., Warren, Z., Kurzius-Spencer, M., Zahorodny, W., Robinson Rosenberg, C., White, T., Durkin, M. S., Imm, P., Nikolaou, L., Yeargin-Allsopp, M., Lee, L., Harrington, R., Lopez, M., Fitzgerald, R. T., Hewitt, A., … Dowling, N. F. (2018). Prevalence of autism spectrum disorder among children aged 8 years—autism and developmental disabilities monitoring network, 11 sites, United States, 2014. *MMWR Surveillance Summaries, 67*(6), 1.

Betancourt, J. R. (2004). Cultural competence—Marginal or mainstream movement? *New England Journal of Medicine, 351*(10), 953–955.

Bhatia-Lin, A., Wong, K., Legha, R., & Walker, V. P. (2021). What will you protect? Redefining professionalism through the lens of diverse personal identities. *MedEdPORTAL, 17*, 11203.

Blau, F. D., & Kahn, L. M. (2017). The gender wage gap: Extent, trends, and explanations. *Journal of Economic Literature, 55*(3), 789–865.

Branch, W. T., Jr. (1998). Professional and moral development in medical students: The ethics of caring for patients. *Transactions of the American Clinical and Climatological Association, 109*, 218.

Brea-Spahn, M. (2021). BLLING learning as belonging. *Perspectives of the ASHA Special Interest Groups, 7*(1), 209–228. https://doi.org/10.1044/leader.AE.26012021.36/full/.

Capers, Q., IV. (2020). How clinicians and educators can mitigate implicit bias in patient care and candidate selection in medical education. *ATS Scholar, 1*(3), 211–217.

Chin, J. L. (2000). Culturally competent health care. *Public Health Reports, 115*(1), 25.

Chinn, J. J., Martin, I. K., & Redmond, N. (2021). Health equity among black women in the United States. *Journal of Women's Health, 30*(2), 212–219.

Dagenais, P. A., & Stallworth, J. A. (2014). The influence of dialect upon the perception of dysarthric speech. *Clinical Linguistics & Phonetics, 28*(7-8), 573–589.

Davis, S. K., Gebreab, S., Quarells, R., & Gibbons, G. H. (2014). Social determinants of cardiovascular health among black and white women residing in Stroke Belt and Buckle regions of the South. *Ethnicity and Disease, 24*(2), 133–143.

Deal-Williams, V. (2010). Coming of age in multicultural affairs. *CSHA Magazine, 40*(1), 10–11.

Deal-Williams, V. R. (2020). Addressing disparities in the wake of injustice, violence, and COVID-19. *Leader Live.* https://leader.pubs.asha.org/do/10.1044/2020-0601-addressing-disparities-of-injustice.

Dee, T. S. (2004). The race connection: Are teachers more effective with students who share their ethnicity? *Education Next, 4*(2), 52–60.

Ding, J., Yong-Hing, C. J., Patlas, M. N., & Khosa, F. (2023). Equity, diversity, and inclusion: Calling, career, or chore? *Canadian Association of Radiologists Journal, 74*(1), 10–11.

Ding, J., Zhou, Y., Khan, M. S., Sy, R. N., & Khosa, F. (2021). Representation of sex, race, and ethnicity in pivotal clinical trials for dermatological drugs. *International Journal of Women's Dermatology, 7*(4), 428–434.

Donini-Lenhoff, F. G., & Hedrick, H. L. (2000). Increasing awareness and implementation of cultural competence principles in health professions education. *Journal of Allied Health, 29*(4), 241–245.

Durán, L. K., Hartzheim, D., Lund, E. M., Simonsmeier, V., & Kohlmeier, T. L. (2016). Bilingual and home language interventions with young dual language learners: A research synthesis. *Language, Speech, and Hearing Services in Schools, 47*(4), 347–371.

Finneran, D. A., Heilmann, J. J., Moyle, M. J., & Chen, S. (2020). An examination of cultural-linguistic influences on PPVT-4 performance in African American and hispanic preschoolers from low-income communities. *Clinical Linguistics & Phonetics, 34*(3), 242–255.

Fish, R. E. (2019). Teacher race and racial disparities in special education. *Remedial and Special Education, 40*(4), 213–224.

Frye, V., Camacho-Rivera, M., Salas-Ramirez, K., Albritton, T., Deen, D., Sohler, N., Barrick, S., & Nunes, J. (2020). Professionalism: The wrong tool to solve the right problem? *Academic Medicine, 95*(6), 860–863.

Fulcher-Rood, K., Castilla-Earls, A. P., & Higginbotham, J. (2018). School-based speech-language pathologists' perspectives on diagnostic decision making. *American Journal of Speech-Language Pathology, 27*(2), 796–812.

Getzler, L., LeMaster, P., & Emerick, M. (2021). Speech language pathologists and teachers perceptions of bilingual students. *Journal of Student Research, 10*(2).

Girolamo, T. M., Politzer-Ahles, S., Ghali, S., & Williams, B. T. (2022). Preliminary evaluation of applicants to master's programs in speech-language pathology using vignettes and criteria from a holistic review process. *American Journal of Speech-Language Pathology, 31*(2), 552–577.

Go, A. S., Mozaffarian, D., Roger, V. L., Benjamin, E. J., Berry, J. D., Blaha, M. J., Dai, S., Ford, E. S., Fox, C. S., Franco, S., Fullerton, H. J., Gillespie, C., Hailpern, S. M., Heit, J. A., Howard, V. J., Huffman, M., Judd, S. E., Kissela, B. M., Kittner, S. J., ... Turner, M. B. (2014). Heart disease and stroke statistics—2014 update: A report from the American Heart Association. *Circulation, 129*(3), e28–e92.

Gray, A. (2019). The bias of "professionalism" standards (SSIR) [Internet]. *Stanford Social Innovation Review*. Retrieved from https://ssir.org/articles/entry/the_bias_of_professionalism_ standards.

Grissom, J. A., Rodriguez, L. A., & Kern, E. C. (2017). Teacher and principal diversity and the representation of students of color in gifted programs: Evidence from national data. *The Elementary School Journal, 117*(3), 396–422.

Guiberson, M., & Vigil, D. C. (2021). Admissions type and cultural competency in graduate speech-language pathology curricula: A national survey study. *American Journal of Speech-Language Pathology, 30*(5), 2017–2027.

Harris, K. M., Gordon-Larsen, P., Chantala, K., & Udry, J. R. (2006). Longitudinal trends in race/ ethnic disparities in leading health indicators from adolescence to young adulthood. *Archives of Pediatrics and Adolescent Medicine, 160*(1), 74–81.

Hill, A. J., Jones, D. B., & Woodworth, L. (2023). Physician-patient race-match reduces patient mortality. *Journal of Health Economics, 92*, 102821.

Holt, S. B., & Gershenson, S. (2015). The impact of teacher demographic representation on student attendance and suspensions. In *IZA discussion papers, No. 9554. Institute for the Study of Labor (IZA), Bonn*.

Hood, S., Hopson, R. K., & Kirkhart, K. (2015). Culturally responsive evaluation: Theory, practice, and future implications. In K. E. Newcomer, H. Hatry, & J. Wholey (Eds.), *Handbook of practical program evaluation* (4th ed., pp. 281–317). San Francisco, CA: Jossey-Bass & Pfeiffer Imprints, Wiley.

Horton, R., Muñoz, M. L., & Johnson, V. E. (2021). Faculty of color, bulletproof souls, and their experiences in communication sciences and disorders. *Perspectives of the ASHA Special Interest Groups, 6*(5), 1227–1244.

Howard, V. J., Cushman, M., Pulley, L., Gomez, C. R., Go, R. C., Prineas, R. J., Graham, A., Moy, C. S., & Howard, G. (2005). The reasons for geographic and racial differences in stroke study: Objectives and design. *Neuroepidemiology, 25*(3), 135–143.

Howard, G., & Howard, V. J. (2020). Twenty years of progress toward understanding the stroke belt. *Stroke, 51*(3), 742–750.

Howard, D. L., Konrad, T. R., Stevens, C., & Porter, C. Q. (2001). Physician-patient racial matching, effectiveness of care, use of services, and patient satisfaction. *Research on Aging, 23*(1), 83–108.

Hyter, Y. D., & Salas-Provance, M. B. (2023). *Culturally responsive practices in speech, language, and hearing sciences* (2nd ed.). San Diego, CA: Plural Publishing.

Jetty, A., Jabbarpour, Y., Pollack, J., Huerto, R., Woo, S., & Petterson, S. (2022). Patient-physician racial concordance associated with improved healthcare use and lower healthcare expenditures in minority populations. *Journal of Racial and Ethnic Health Disparities*, 1–4.

Jones, E. P., Margolius, M., Rollock, M., Yan, C. T., Cole, M. L., & Zaff, J. F. (2018). Disciplined and disconnected: How students experience exclusionary discipline in Minnesota and the promise of non-exclusionary alternatives. *America's Promise Alliance*. Retrieved March 1, 2023 from https://files.eric.ed.gov/fulltext/ED586336.pdf.

Kasambira Fannin, D. K. (2020). Celebrating the 50th anniversary of ASHA's office of multicultural affairs. *Perspectives of the ASHA Special Interest Groups, 5*(1), 1–2.

Kasambira Fannin, D. K., & Mandulak, K. C. (2021). Introduction to the forum: Increasing diversity in the communication sciences and disorders workforce, part 1. *American Journal of Speech-Language Pathology, 30*(5), 1913–1915.

Kim, E. T., Franz, L., Kasambira Fannin, D., Howard, J., & Maslow, G. (2021). Educational classifications of autism spectrum disorder and intellectual disability among school-aged children in North Carolina: Associations with race, rurality, and resource availability. *Autism Research, 14*(5), 1046–1060.

Kritikos, E. P. (2003). Speech-language pathologists' beliefs about language assessment of bilingual/bicultural individuals. *American Journal of Speech-Language Pathology, 12*(1), 73–91.

Lackland, D. T., Roccella, E. J., Deutsch, A. F., Fornage, M., George, M. G., Howard, G., Kissela, B. M., Kittner, S. J., Lichtman, J. H., Lisabeth, L. D., & Schwamm, L. H. (2014). Factors influencing the decline in stroke mortality: A statement from the American Heart Association/American Stroke Association. *Stroke, 45*(1), 315–353.

Langford-Hall, M. (2016). Recruitment, retention and mentoring of minorities into the fields of communication sciences and disorders. *International Journal of Humanities and Social Science Review, 2*(9), 1–4.

Magaña, S., Parish, S. L., Rose, R. A., Timberlake, M., & Swaine, J. G. (2012). Racial and ethnic disparities in quality of health care among children with autism and other developmental disabilities. *Intellectual and Developmental Disabilities, 50*(4), 287–299.

Mahendra, N., Girolamo, T. M., & Kasambira Fannin, D. (2021). Advancing justice, equity in the pipeline to the professions: Reconfiguring graduate training program admissions is key to foundational change. *The ASHA Leader, 26*(6), 8.

Mahendra, N., & Spicer, J. (2014). Access to speech-language pathology services for African-American clients with aphasia: A qualitative study. *Division 14 Newsletter, 21*(2), 53–62.

Mahendra, N., & Visconti, C. F. (2021). Racism, equity and inclusion in communication sciences and disorders: Reflections and the road ahead. *Teaching and Learning in Communication Sciences and Disorders, 5*(3), 1.

Mandulak, K. C. (2022). The case for holistic review in communication sciences and disorders admissions. *Perspectives of the ASHA Special Interest Groups, 7*(2), 476–481.

Mandulak, K. C., & Kasambira Fannin, D. (2023, April 12). Admissions summit [workshop]. In *CAPCSD 43rd annual conference, Orlando, FL.*

Matthews, J. J., & Daniels, D. E. (2019). The gendered experiences of male students in a speech-language pathology graduate program: A multi-case study. *Teaching and Learning in Communication Sciences and Disorders, 3*(2), 2.

Morgan, P. L., Farkas, G., Cook, M., Strassfeld, N. M., Hillemeier, M. M., Pun, W. H., & Schussler, D. L. (2017). Are black children disproportionately overrepresented in special education? A best-evidence synthesis. *Exceptional Children, 83*(2), 181–198.

Mozaffarian, D., Benjamin, E. J., Go, A. S., Arnett, D. K., Blaha, M. J., Cushman, M., Das, S. R., Ferranti, J. D., Fullerton, H. J., Howard, V. J., Huffman, M. D., Isasi, C. R., Jiménez, M. C., Judd, S. E., Kissela, B. M., Lichtman, J. H., Lisabeth, L. D., Liu, S., … Turner, M. B. (2016). Heart disease and stroke statistics—2016 update: A report from the American Heart Association. *Circulation, 133*(4), e38–e60.

Myers, K., Trull, L. H., Bryson, B. J., & Yeom, H. S. (2019). Affinity groups: Redefining brave spaces. *Journal of Baccalaureate Social Work, 24*(1), 1–8.

Narayanan, T. L., & Ramsdell, H. L. (2022). Self-reported confidence and knowledge-based differences between multilingual and monolingual speech-language pathologists when serving culturally and linguistically diverse populations. *Perspectives of the ASHA Special Interest Groups, 7*(1), 209–228.

National Center for Education Statistics. (2015). *Postsecondary attainment: Differences by socioeconomic status*. Retrieved May 12, 2023 from https://nces.ed.gov/programs/coe/pdf/coe_tva.pdf.

National Center for Education Statistics. (2022). Higher Education General Information Survey (HEGIS), Fall Enrollment in Colleges and Universities, 1976 through 1985 surveys. In *Integrated Postsecondary Education Data System (IPEDS), "Fall Enrollment Survey" (IPEDS-EF: 86-99); and IPEDS Spring 2001 through Spring 2022, Fall Enrollment component*. Retrieved May 17, 2023 from https://nces.ed.gov/programs/digest/d22/tables/dt22_313.20.asp.

National Center for Education Statistics. (2023). *Students With disabilities*. U.S. Department of Education, Institute of Education Sciences. Retrieved April 6, 2023 from https://nces.ed.gov/programs/coe/indicator/cgg.

National Center for Learning Disabilities. (2022). *IDEA full funding: Why should congress invest in special education?*. Retrieved March 6, 2023 from https://ncld.org/news/policy-and-advocacy/idea-full-funding-why-should-congress-invest-in-special-education/.

New York University. (2016). Race influences teachers' referrals to special and gifted education, finds study. *ScienceDaily*. Retrieved March 3, 2023 from http://www.sciencedaily.com/releases/2016/10/161018094738.htm.

Newkirk-Turner, B. L., & Hudson, T. K. (2022). Do no harm: Graduate admissions letters of recommendation and unconscious bias. *Perspectives of the ASHA Special Interest Groups*, 7(2), 463–475.

Newman-Toker, D. E. (2014). A unified conceptual model for diagnostic errors: Underdiagnosis, overdiagnosis, and misdiagnosis. *Diagnosis*, 1(1), 43–48.

Norman, G. R., & Eva, K. W. (2010). Diagnostic error and clinical reasoning. *Medical Education*, 44(1), 94–100.

Núñez, G., & Hughes, M. T. (2021). Transforming language intervention: Collaborating with Latinx families. *Childhood Education*, 97(3), 54–59.

Office of Special Education and Rehabilitative Services. (2021). *Racial and ethnic disparities in special education: A multi-year disproportionality analysis by state, analysis category, and race/ethnicity*. US Department of Education (ED). Retrieved April 6, 2023 from https://www2.ed.gov/programs/osepidea/618-data/LEA-racial-ethnic-disparities-tables/.

Office of Special Education Programs. (2021a). *OSEP fast facts: Race and ethnicity of children with disabilities served under IDEA Part B*. Retrieved January 5, 2023 from https://sites.ed.gov/osers/2021/08/osep-releases-fast-facts-on-the-race-and-ethnicity-of-children-with-disabilities-served-under-idea-part-b/.

Office of Special Education Programs. (2021b). *Fast facts: Historically black colleges and universities*. Retrieved January 5, 2023 from https://sites.ed.gov/osers/2021/08/osep-releases-fast-facts-on-the-race-and-ethnicity-of-children-with-disabilities-served-under-idea-part-b/.

O'Malley, K. (2015). From mainstreaming to marginalization-IDEA's de facto segregation consequences and prospects for restoring equity in special education. *University of Richmond Law Review*, 50, 951.

Ouazad, A. (2014). Assessed by a teacher like me: Race and teacher assessments. *Education Finance and Policy*, 9(3), 334–372.

Parveen, S., & Santhanam, S. P. (2021). Speech-language pathologists' perceived competence in working with culturally and linguistically diverse clients in the United States. *Communication Disorders Quarterly*, 42(3), 166–176.

Paz, S., Alfano, A. R., Medina, A. M., & Hayes, T. (2023). Speech-language pathologists' perceptions of childhood bilingualism. *Clinical Linguistics and Phonetics, 1-20.* https://doi.org/10.1080/02699206.2023.2021204.

Penner, L. A., Dovidio, J. F., West, T. V., Gaertner, S. L., Albrecht, T. L., Dailey, R. K., & Markova, T. (2010). Aversive racism and medical interactions with black patients: A field study. *Journal of Experimental Social Psychology, 46*(2), 436–440.

Perrachione, T. K., Chiao, J. Y., & Wong, P. C. (2010). Asymmetric cultural effects on perceptual expertise underlie an own-race bias for voices. *Cognition, 114*(1), 42–55.

Petersen, S. L., Erenrich, E. S., Levine, D. L., Vigoreaux, J., & Gile, K. (2018). Multi-institutional study of GRE scores as predictors of STEM PhD degree completion: GRE gets a low mark. *PLoS One, 13*(10), e0206570.

Porter, S. (1993). Critical realist ethnography: The case of racism and professionalism in a medical setting. *Sociology, 27*(4), 591–609.

Richburg, C. M. (2022). Underrepresentation of students from diverse backgrounds entering communication sciences and disorders programs: An investigation into the university student perspective. *American Journal of Speech-Language Pathology, 31*(2), 613–630.

Robinson, G. C., & Norton, P. C. (2019). A decade of disproportionality: A state-level analysis of African American students enrolled in the primary disability category of speech or language impairment. *Language, Speech, and Hearing Services in Schools, 50*(2), 267–282.

Rogus-Pulia, N., Humbert, I., Kolehmainen, C., & Carnes, M. (2018). How gender stereotypes may limit female faculty advancement in communication sciences and disorders. *American Journal of Speech-Language Pathology, 27*(4), 1598–1611.

Salsberg, E., Richwine, C., Westergaard, S., Martinez, M. P., Oyeyemi, T., Vichare, A., & Chen, C. P. (2021). Estimation and comparison of current and future racial/ethnic representation in the US health care workforce. *JAMA Network Open, 4*(3), e213789.

Schaeffer, K. (2021). *America's public school teachers are far less racially and ethnically diverse than their students.* Retrieved March 1, 2023 from https://www.pewresearch.org/short-reads/2021/12/10/americas-public-school-teachers-are-far-less-racially-and-ethnically-diverse-than-their-students/.

Scott, R. D. (2023). Minnesota CSD program formalizes support for students of color. *Perspectives of the ASHA Special Interest Groups, 7*(1), 209–228.

Sealy, L., Saunders, C., Blume, J., & Chalkley, R. (2019). The GRE over the entire range of scores lacks predictive ability for PhD outcomes in the biomedical sciences. *PLoS One, 14*(3), e0201634.

Shipley, K. G., & McAfee, J. G. (2019). *Assessment in speech-language pathology: A resource manual.* San Diego, CA: Plural Publishing.

Stanford, S. (2019). Casualties of misunderstanding: Communication disorders and juvenile injustice. *The ASHA Leader, 24*(6), 44–54.

Stanford, S. (2020). The school-based speech-language pathologist's role in diverting the school-to-confinement pipeline for youth with communication disorders. *Perspectives of the ASHA Special Interest Groups, 5*(4), 1057–1066.

Stockman, I. J., Boult, J., & Robinson, G. C. (2008). Multicultural/multilingual instruction in educational programs: A survey of perceived faculty practices and outcomes. *American Journal of Speech-Language Pathology, 17*(3), 241–264.

Tao, B. K., Vosoughi, A. R., He, B., Ling, J., Xia, M., Rocha, G., Ing, E., & Khosa, F. (2023). Representational disparity of sex, race, and ethnicity in presbyopia clinical trials: a cross-sectional study. *Eye, 37*(18), 3871–3873.

Thompson, D. (2013, November 6). *The 33 whitest jobs in America*. The Atlantic.

U.S. Census Bureau. (2020). *Quickfacts: United States*. Retrieved May 6, 2023 from https://www. census.gov/quickfacts/fact/table/US/PST045222.

U.S. Department of Health and Human Services. (2022). *Social determinants of health*. Retrieved January 4, 2023 from https://health.gov/healthypeople/priority-areas/social-determinants-health.

Watts, C. R., DiLollo, A., & Zhang, Y. (2023). The impact of academic, sociodemographic, and program growth factors on admission offers to US graduate education programs in communication sciences and disorders: National trends in 2016–2020 cycles. *American Journal of Speech-Language Pathology*, *32*(1), 1–2.

Winstein, C. J., Stein, J., Arena, R., Bates, B., Cherney, L. R., Cramer, S. C., Deruyter, F., Eng, J. J., Fisher, B., Harvey, R. L., Lang, C. E., MacKay-Lyons, M., Ottenbacher, K. J., Pugh, S., Reeves, M. J., Richards, L. G., Stiers, W., & Zorowitz, R. D. (2016). Guidelines for adult stroke rehabilitation and recovery: A guideline for healthcare professionals from the American Heart Association/American Stroke Association. *Stroke*, *47*(6), e98–e169.

Wong, A. A., Marrone, N. L., Fabiano-Smith, L., Beeson, P. M., Franco, M. A., Subbian, V., & Lozano, G. I. (2021). Engaging faculty in shifting toward holistic review: Changing graduate admissions procedures at a land-grant, Hispanic-serving institution. *American Journal of Speech-Language Pathology*, *30*(5), 1925–1939.

Woods, M. (2022). *Communication sciences and disorders centralized application service 2021-2022 applicant data report: 2021-2022 admissions cycle for the 2022 entering class*. Retrieved April 20, 2023 from https://growthzonesitesprod.azureedge.net/wp-content/uploads/sites/1023/2023/04/2021-2022-CSDCAS-Applicant-Data-Report-.pdf.

Yingling, M. E., Hock, R. M., & Bell, B. A. (2018). Time-lag between diagnosis of autism spectrum disorder and onset of publicly-funded early intensive behavioral intervention: Do race–ethnicity and neighborhood matter? *Journal of Autism and Developmental Disorders*, *48*, 561–571.

Yong-Hing, C. J., & Khosa, F. (2023). Provision of culturally competent healthcare to address healthcare disparities. *Canadian Association of Radiologists Journal*, *74*(3), 483–484.

Yong-Hing, C. J., Vaqar, M., Sahi, Q., & Khosa, F. (2023). Burnout: Turning a crisis into an opportunity. *Canadian Association of Radiologists Journal*, *74*(1), 16–17.

Young, P. J., Kagetsu, N. J., Tomblinson, C. M., Snyder, E. J., Church, A. L., Mercado, C. L., Guzman Perez-Carillo, G. J., Jha, P., Guerro-Calderon, J. D., Jaswal, S., Khosa, F., & Deitte, L. A. (2023). The intersection of diversity and well-being. *Academic Radiology*, *30*(9), 2031–2036.

Yu, B., Horton, R., Munson, B., Newkirk-Turner, B. L., Johnson, V. E., Khamis-Dakwar, R., Muñoz, M. L., & Hyter, Y. D. (2022). Making race visible in the speech, language, and hearing sciences: A critical discourse analysis. *American Journal of Speech-Language Pathology*, *31*(2), 578–600.

Conclusion: The road ahead

Faisal Khosa, Jeffrey Ding, and Sabeen Tiwana
University of British Columbia, Vancouver, BC, Canada

Even if you're on the right track, you'll get run over if you just sit there.

Will Rogers

Equity, diversity, and inclusion (EDI) is an evolving concept that continues to expand its scope of work. As demonstrated in the various chapters, the different healthcare professions each face unique challenges in representation and workforce inequity. It is important to recognize that inequities fluctuate as the population demographics shift over time; therefore, there must be constant reevaluation and surveillance of the state of EDI in the workforce pipeline. Furthermore, it is imperative to uphold the principles of cultural safety, cultural humility, and cultural competency when engaging in EDI work. Cultural safety is an outcome produced through addressing power imbalances and facilitating respectful engagement. This in turn supports a safe environment free of racism and discrimination. Cultural humility is a process that develops relationships based on mutual trust. It refers to the process of identifying self- and systemic-conditioned biases, which are exhibited by all individuals irrespective of their background. Lastly, cultural competency is the awareness that the process is more important than the product—to become fully competent in another person's culture is not the goal.

There are tangible and intangible benefits of EDI including enhanced team performance, improved employee morale, reduced burnout, employee retention and engagement, as well as increased profitability and success (Wei et al., 2023; Yong-Hing et al., 2023; Yong-Hing & Khosa, 2023; Young et al., 2023). EDI in teams also provides mentors and role models for students and trainees from minority groups (Deanna et al., 2022). Teams with diverse perspectives have enhanced problem-solving capabilities and cultural humility, allowing for the provision of culturally competent care leading to enhanced patient satisfaction and better outcomes (Stubbe, 2020). Furthermore, research has shown that when the principal investigator of a clinical trial is a female, the patient population that is enrolled in the study is more equitable (Yong et al., 2023). It is imperative that the study populations represented within clinical trials equitably represent the demographics of the general population to ensure that any safety and efficacy data is obtained from representative sampling (Ding et al., 2021, 2022).

While the EDI movement has gained traction over recent years, disparities remain widespread and deep-seated across the healthcare systems. Non-White communities experience adverse health outcomes in the form of receiving lower quality health services compared to their White counterparts (Smedley et al., 2003). Indigenous populations in the United States (US) are profoundly impacted as they have a lower life expectancy of more than 5 years compared to non-Indigenous individuals (Allan & Smylie, 2015; Reading & Wien, 2009). Similar patterns in life expectancy differences have been noted in Canada as well (Healthy aging report, 2017). Pertaining to sexism, reduced healthcare access for women is a driving factor for gender disparity in health outcomes; however, it is important to note that this further intersects with other influences such as race, class, and ability (Doyal, 1995; Macintyre et al., 1996). As for transgender individuals, the consequences of systems of oppression upholding cisnormativity and perpetuating transphobia can be seen in an example from Canada, where a study found that 10% of trans people were either refused care or received incomplete care from the emergency department and 40% were subjected to discriminatory behavior in the primary care setting (Bauer & Scheim, 2015). Individuals who are lesbian, bisexual, or gay experience similar health inequities fueled by homophobia and heteronormativity (Fredriksen-Goldsen et al., 2013; Meyer, 2003). Lastly, ableism such as disability-based discrimination interacts concurrently with all of the above to further accentuate poor health outcomes and drivers of inequity (Emerson et al., 2011; Krnjacki et al., 2018).

Understanding the unearned advantages/disadvantages and putting one's privilege into perspective is an important step to embracing EDI. The "privilege walk" is a self and/or team reflection activity proposed by Jennifer Brown, which serves to identify privilege as it relates to gender, race, class, sexual orientation, and more (Brown, 2019). Participants are asked a series of questions, and one's answers dictate whether a step should be taken forward or backward. At the end of this activity, the degree of privilege of an individual can be gauged by how many steps forward they have taken in relation to the others, all of whom began at the same starting line. For example, individuals are asked to take a step forward if they were born in the US, went to college, if they can show affection toward their romantic partner in public without fear of ridicule, or if they feel comfortable walking home alone at night, among others. Meanwhile, individuals are asked to take a step back if they took loans for their education, if they have ever been diagnosed as having a disability, or if they have ever been the only person of their gender/socioeconomic status/race/sexual orientation in a workplace setting, among others (Brown, 2019). Coming to terms with our privileges, or lack thereof, is a crucial step to understanding an individual's role in systems of oppression.

The value of mentorship is often brought up in the discussion of fostering greater EDI in the healthcare professions. By aligning oneself with mentors, role models, and same-level peers, one's overall career trajectory can benefit greatly through these win-win mentor-mentee relationships. However, while the benefits of mentorship and sponsorship are widely documented, individuals from marginalized backgrounds may not always be supportive of one another. For example, research has shown that some women in established positions of leadership have been noted to distance themselves or exhibit an aversion to aligning too closely with other women, as a way of maintaining their competitiveness and avoiding impediments to their continued career advancement (Galsanjigmed & Sekiguchi, 2023). This dilemma is prevalent across the healthcare professions. Education and healthcare professions by design promote a win-lose mindset. From the start (e.g., applying for admission into a training program), one must be a competitive applicant to obtain acceptance—for one person to succeed, several others must fail (i.e., rejection from the program). The challenge is that once in professional practice, there is a sudden expectation for proficiency in the win-win mindset—the way of thinking goes from "how do *I* become successful" to "how do *we ensure team* success."

It is imperative that we embrace the win-win mindset, while recognizing the learning curve that we must overcome when adopting this way of collaborative practice.

For those in positions of influence and leadership who hope to nurture and cultivate the next generation of leaders, the art of providing feedback is something that requires careful attention. Feedback can be counterintuitive when it is not tactfully crafted, and in other cases, it can be intentionally harmful. Biased feedback with undertones of stereotypes is common toward women, meanwhile Asian women are referred to as "submissive" or "docile," and Black women are dubbed "aggressive" (Pay Women of Color Equally for Equal Work, 2022). The Situation-Behavior-Impact feedback framework is helpful in providing on-the-spot feedback by addressing precise behavior that one would like to comment on and specifically identifying the impact of the behavior on an individual, team, or organization (Use Situation-Behavior-Impact (SBI)™ to understand intent, 2022). Furthermore, understanding the ramifications and implications of one's choice of words is also critical in delivering effective and constructive feedback. For instance, careful attention to terms such as "angry" (which could be substituted with "passionate"), "bossy" (as opposed to "leader-like"), or hostile (versus "confident").

For anyone in the healthcare workforce, regardless of whether or not they are in a position of leadership, it can be helpful to critically appraise one's work environment. Complacency toward discriminatory or toxic work environments is a threat to EDI in the workplace. Perpetuating harmful workplace practices can be especially detrimental to members from marginalized groups. Unfair policies or practices do not adversely affect all individuals equally, but all individuals have the right to challenge the existing paradigm and bring about change. As Henry Ford said, "If you always do what you've always done, you always get what you've always gotten." In *DEI for Dummies*, it is suggested that you can evaluate qualities of one's workplace and identify possible toxic qualities by looking at five aspects: (1) communication (e.g., passive-aggressive messages, unclear or vague work assignments), (2) career opportunities (e.g., rare promotions, no clear path to grow), (3) leadership (e.g., micromanaged, lack of positive reinforcement, lack of representation within leadership), (4) peers and co-workers (e.g., unmotivated colleagues, exclusive cliques within the team, widespread gossip), and (5) work-life balance (e.g., people rarely take vacations, feeling proud when working excessive hours) (Davis, 2021).

The compounding obstacles to EDI in the workforce pipeline have been alluded to the "sticky floor, broken ladder, and glass ceiling" (Kim et al., 2022; Lail et al., 2024; Maqsood et al., 2021). The metaphor for impediments within institutions which, by design, hinder the progress of the underrepresented demographics is the "sticky floor." Additionally, the "broken ladder" refers to the inequitable process by which leadership positions are transferred. When a job is advertised in a way to cater to certain individuals, there are members of certain demographic groups who are inherently disadvantaged. A prevailing institutional culture is that a candidate is chosen before the job is advertised, and hence certain individuals among select demographics are unfairly favored. Furthermore, the "glass ceiling" comments on the self-perception, commonly by members from underrepresented groups (such as but not limited to women and people of color), of being underqualified or overqualified, yet never appropriately qualified for opportunities. Sticky floor, broken ladder, and glass ceiling interact with one another in an additive and compounding manner, which effectively reinforces disparities in the workforce (Kim et al., 2022).

In conclusion, our collaboration on EDI in the healthcare profession sheds light on the value of fostering a more equitable healthcare workforce. While the journey through the pages of this text comes to an end, the act of cultivating and supporting EDI is not a destination but an ongoing process. Some healthcare professions are well on their way to promoting greater equity, meanwhile others have just begun—the message is that there is space for EDI in all realms of healthcare and, irrespective of background or identity, there is always a role to be played whether it is from someone with privilege or unearned disadvantage. Regardless of one's gender, race, sexual orientation, age, visible or invisible disability, and many other demographic classifications, we eagerly welcome everyone to join in our shared pursuit to create a more equitable and compassionate healthcare system. As we reach the final pages, it is important to remember that here only marks the beginning of an era of greater EDI. We extend our heartfelt gratitude to all the authors, readers, practitioners, and advocates who have embarked on this transformative adventure with us. The editors welcome feedback on how to enhance and improve the current content in future editions of this work.

Justice must not only be done but must manifestly and undoubtedly be seen to be done.

Lord Chief Justice Hewart

References

Allan, B., & Smylie, J. (2015). *First peoples, second class treatment: The role of racism in the health and well-being of Indigenous peoples in Canada.* Wellesley Institute. https://www.wellesleyinstitute.com/wp-content/uploads/2015/02/Summary-First-Peoples-Second-Class-Treatment-Final.pdf.

Bauer, G. R., & Scheim, A. I. (2015). *Transgender people in Ontario, Canada: Statistics from the trans PULSE project to inform human rights policy.* TransPULSE. https://transpulseproject.ca/wp-content/uploads/2015/06/Trans-PULSE-Statistics-Relevant-for-Human-Rights-Policy-June-2015.pdf.

Brown, J. (2019). *How to be an inclusive leader: Your role in creating cultures of belonging where everyone can thrive* (1st ed.). Berrett-Koehler Publishers, Inc.

Davis, S. (2021). *Diversity, equity, and inclusion for dummies* (1st ed.). John Wiley and Sons.

Deanna, R., Merkle, B. G., Chun, K. P., Navarro-Rosenblatt, D., Baxter, I., Oleas, N., Bortolus, A., Geesink, P., Diele-Viegas, L., Aschero, V., De Leone, M. J., Oliferuk, S., Zuo, R., Cosacov, A., Grossi, M., Knapp, S., Lopez-Mendez, A., Welchen, E., Ribone, P., & Auge, G. (2022). Community voices: The importance of diverse networks in academic mentoring. *Nature Communications, 13*(1), 1681. https://doi.org/10.1038/s41467-022-28667-0.

Ding, J., Haq, A. F., Joseph, M., & Khosa, F. (2021). Disparities in pediatric clinical trials for acne vulgaris: A cross-sectional study. *Journal of the American Academy of Dermatology,* S0190962221026542. https://doi.org/10.1016/j.jaad.2021.10.013.

Ding, J., Joseph, M., Chawla, S., Yau, N., & Khosa, F. (2022). Disparities in alopecia clinical trials: An analysis of female and minority representation. *Journal of Cutaneous Medicine and Surgery,* 12034754221099667. https://doi.org/10.1177/12034754221099667.

Doyal, L. (1995). *What makes women sick.* Macmillan Education UK. https://doi.org/10.1007/978-1-349-24030-2.

Emerson, E., Madden, R., Graham, H., Llewellyn, G., Hatton, C., & Robertson, J. (2011). The health of disabled people and the social determinants of health. *Public Health, 125*(3), 145–147. https://doi.org/10.1016/j.puhe.2010.11.003.

Fredriksen-Goldsen, K. I., Kim, H.-J., Barkan, S. E., Muraco, A., & Hoy-Ellis, C. P. (2013). Health disparities among lesbian, gay, and bisexual older adults: Results from a population-based study. *American Journal of Public Health, 103*(10), 1802–1809. https://doi.org/10.2105/AJPH.2012.301110.

Galsanjigmed, E., & Sekiguchi, T. (2023). Challenges women experience in leadership careers: An integrative review. *Merits, 3*(2), 366–389. https://doi.org/10.3390/merits3020021.

Healthy aging report. (2017). Native Women's Association of Canada. https://www.nwac.ca/assets-knowledge-centre/Healthy-Aging-Report.pdf.

Kim, K. Y., Kearsley, E. L., Yang, H. Y., Walsh, J. P., Jain, M., Hopkins, L., Wazzan, A. B., & Khosa, F. (2022). Sticky floor, broken ladder, and glass ceiling in academic obstetrics and gynecology in the United States and Canada. *Cureus.* https://doi.org/10.7759/cureus.22535.

Krnjacki, L., Priest, N., Aitken, Z., Emerson, E., Llewellyn, G., King, T., & Kavanagh, A. (2018). Disability-based discrimination and health: Findings from an Australian-based population study. *Australian and New Zealand Journal of Public Health, 42*(2), 172–174. https://doi.org/10.1111/1753-6405.12735.

Lail, A., Ding, J., Leyva, B. K., Jalal, S., Nakae, S., Fares, S., & Khosa, F. (2024). Ivory tower in MD/PhD programmes: sticky floor, broken ladder and glass ceiling. *BMJ Leader,* leader-2024-001003. https://doi.org/10.1136/leader-2024-001003.

Macintyre, S., Hunt, K., & Sweeting, H. (1996). Gender differences in health: Are things really as simple as they seem? *Social Science & Medicine, 42*(4), 617–624. https://doi.org/10.1016/0277-9536(95)00335-5.

Maqsood, H., Younus, S., Naveed, S., Chaudhary, A. M. D., Khan, M. T., & Khosa, F. (2021). Sticky floor, broken ladder, and glass ceiling: Gender and racial trends among neurosurgery residents. *Cureus, 13*(9), e18229. https://doi.org/10.7759/cureus.18229.

Meyer, I. H. (2003). Prejudice, social stress, and mental health in lesbian, gay, and bisexual populations: Conceptual issues and research evidence. *Psychological Bulletin, 129*(5), 674–697. https://doi.org/10.1037/0033-2909.129.5.674.

Pay Women of Color Equally for Equal Work. (2022). In R. Tulshyan (Ed.), *Inclusion on purpose: An intersectional approach to creating a culture of belonging at work* (pp. 121–142). The MIT Press. https://doi.org/10.7551/mitpress/14004.003.0013.

Reading, C., & Wien, F. (2009). *Health inequalities and social determinants of aboriginal peoples' health.* National Collaborating Centre for Aboriginal Health. https://www.ccnsa-nccah.ca/docs/determinants/RPT-HealthInequalities-Reading-Wien-EN.pdf.

Smedley, B. D., Stith, A. Y., & Nelson, A. R. (Eds.). (2003). *Unequal treatment: Confronting racial and ethnic disparities in health care (with CD)* (p. 12875). National Academies Press. https://doi.org/10.17226/12875.

Stubbe, D. E. (2020). Practicing cultural competence and cultural humility in the care of diverse patients. *Focus, 18*(1), 49–51. https://doi.org/10.1176/appi.focus.20190041.

Use Situation-Behavior-Impact (SBI)™ to understand intent. (2022, November 18). Center for Creative Leadership. https://www.ccl.org/articles/leading-effectively-articles/closing-the-gap-between-intent-vs-impact-sbii/.

Wei, W., Cai, Z., Ding, J., Fares, S., Patel, A., & Khosa, F. (2023). Organizational leadership gender differences in medical schools and affiliated universities. *Journal of Women's Health*, jwh.2023.0326. https://doi.org/10.1089/jwh.2023.0326.

Yong, C., Suvarna, A., Harrington, R., Gummidipundi, S., Krumholz, H. M., Mehran, R., & Heidenreich, P. (2023). Temporal trends in gender of principal investigators and patients in cardiovascular clinical trials. *Journal of the American College of Cardiology, 81*(4), 428–430. https://doi.org/10.1016/j.jacc.2022.10.038.

Yong-Hing, C. J., & Khosa, F. (2023). Provision of culturally competent healthcare to address healthcare disparities. *Canadian Association of Radiologists Journal, 74*(3), 483–484. https://doi.org/10.1177/08465371231154231.

Yong-Hing, C. J., Vaqar, M., Sahi, Q., & Khosa, F. (2023). Burnout: Turning a crisis into an opportunity. *Canadian Association of Radiologists Journal, 74*(1), 16–17. https://doi.org/10.1177/08465371221130683.

Young, P. J., Kagetsu, N. J., Tomblinson, C. M., Snyder, E. J., Church, A. L., Mercado, C. L., Guzman Perez-Carrillo, G. J., Jha, P., Guerrero-Calderon, J. D., Jaswal, S., Khosa, F., & Deitte, L. A. (2023). The intersection of diversity and well-being. *Academic Radiology, 30*(9), 2031–2036. https://doi.org/10.1016/j.acra.2023.01.028.

Index

Note: Page numbers followed by *f* indicate figures.